Indiana Slavic Studies

VOLUME IV

Indiana Slavic Studies

VOLUME IV

Editor

WILLIAM B. EDGERTON

Associate Editors

WALTER N. VICKERY

JOHN FRED BEEBE

1967
MOUTON
THE HAGUE · PARIS

Published simultaneously in the United States of America by the Indiana University Press and in The Netherlands by Mouton & Co.

Printed in The Netherlands by Mouton & Co., Printers, The Hague.

TABLE OF CONTENTS

FORMULAIC STYLE AND THE RUSSIAN BYLINA

PATRICIA ARANT

Brown University

"It's too late to correct it", said the Red Queen: "when you've once said a thing, that fixes it, and you must take the consequences."

I

AIM

The concept of the formulaic character of oral epic poetry and the usefulness of the formulaic method of composition to the oral singer in the rapid composition of his tales is no longer a new one.[1] A technique of formula analysis already has been carefully worked out for the Yugoslav and Homeric oral epic traditions by Albert Bates Lord,[2] who has elaborated a theory first formulated by Milman Parry in the thirties of this century.[3] This theory centers around the idea that a poet in an oral epic tradition composes his poems in performance, and because of this a special technique of composition is required for the rapid telling of his tales. Francis P. Magoun, Jr. and Robert Creed have subjected the Old English *Beowulf* to this type of analysis with successful results.[4] In addition, the *chansons de gestes*, the Spanish ballad and Gaelic oral poetry also have received similar treatment.[5] This technique of formula analysis now will be applied to a

[1] Preparation of this monograph was made possible by grants from the Committee on the Humanities and the Committee on Summer Grants of Vanderbilt University. This study is based, in part, on my unpublished dissertation, "Compositional Techniques of the Russian Oral Epic, the *Bylina*" (Harvard, 1963).

[2] Albert B. Lord, *The Singer of Tales* (Cambridge, Mass., 1960).

[3] Milman Parry, "Studies in the Epic Technique of Oral Verse-Making. I: Homer and Homeric Style", *Harvard Studies in Classical Philology*, XLI (1930), 73-147.

[4] Francis P. Magoun, "Oral-Formulaic Character of Anglo-Saxon Narrative Poetry", *Speculum*, XXVIII (July 1953), 446-467; Robert P. Creed, "Studies in the techniques of composition of the Béowulf poetry in British Museum Ms. Cotton Vitellius A. XV" (unpubl. diss. Harvard, 1956). See also Lewis E. Nicholson, "Oral Techniques in the Composition of Expanded Anglo-Saxon Verses", *PMLA*, LXXVIII (Sept. 1963), 287-292.

[5] Jean Rychner, *La Chanson de Geste, Essai sur l'Art Épique des Jongleurs* (Geneva-Lille, 1955); Stephen G. Nichols, Jr., *Formulaic Diction and Thematic Composition in the Chanson de Roland* (= *University of North Carolina Studies in the Romance Languages and Literatures*,

selected corpus of Russian material in an attempt to discover, in the first place, whether Parry and Lord's theory and method, which were so successfully applied to the epic songs of one Slavic oral tradition, the Yugoslav, can be applied with equal success to the epic songs of the Russian oral tradition, and to see, in the second place, whether any additional information can be obtained from the Russian material, if the theory and method are applicable.[6] The question, in other words, is whether a similar investigation of the Russian material could have led Parry and Lord to the same conclusions reached from their study of the Yugoslav oral epic tradition, which submitted easily to this kind of analysis.

Study of the special technique that makes it possible for an oral poet to compose in performance will be concerned primarily with form, as many other studies of the *bylina* have been, especially those made both in Russia and abroad within the past thirty years or so. The focus will be shifted, however, to those particular aspects of form that will help us understand the special way in which an oral poet in the Russian epic tradition composes his songs. The scope of this study, therefore, is limited to the form and manner of composition of the metrical line. This approach proceeds from the belief that scholars who have studied Russian oral epic song thus far have been content to think "in terms of a fixed text or a fixed group of texts to which a poet *did* something for a reason within his own artistic or intellectual self", at the same time failing to take into account the possibility that a poet composes in a particular way "because of necessity or because of the demands of his traditional art".[7]

II

SOME EARLIER DISCUSSIONS

Studies made of the Russian *bylina* are endless in number.[8] At our disposal is a valuable corpus of research from the mythological, comparative, historical,

No. 36, Chapel Hill, 1961); Ruth House Webber, *Formulistic Diction in the Spanish Ballad*, University of California Publications in Modern Philology, XXXIV (Berkeley, 1951), 175-278; James Ross, "Formulaic Composition in Gaelic Oral Literature", *Modern Philology*, LVII (Aug. 1959), 1-12.

[6] See Elizabeth Stenbock-Fermor, rev. of A. B. Lord, *The Singer of Tales* (Cambridge, Mass., 1960), in *The Slavic and East European Journal*, V (Summer 1961), 159-161. In this review Mrs. Stenbock-Fermor suggests that it would be useful to apply formula analysis to Russian oral epic verse.

[7] Lord, *op. cit.*, p. 11.

[8] For a survey of the work done in the field of folklore in Soviet Russia see Felix J. Oinas, "Folklore Activities in Russia", *Journal of American Folklore*, LXXIV (1961), 362-370. Elements of a complete bibliography of folklore studies made in Russia during both the Soviet and prerevolutionary periods can be found in this article.

sociological and other points of view outside the realm of present interest. Certain formal aspects of direct or indirect concern in this study also have been frequently and thoroughly treated by scholars. None of these studies, however, gives us a clear picture of how the poet in this oral tradition composes his songs.

Interest in the collection and publication of *byliny*, and of folklore in general, arose in Russia in the nineteenth century in connection with the Romantic movement, receiving special inspiration from Germany and the brothers Grimm. In the second half of the nineteenth century this interest increased significantly due to the discovery of a living *bylina* tradition in the Lake Onega region of Karelia by P. N. Rybnikov. At this point, Russian collectors and scholars began to turn their attention to the individual singer, to his creative personality, and to his repertory and style. This approach, among other things, called attention to the fact that the content of a *bylina* from performance to performance was unstable. As early as the 1860's Rybnikov, in his "Zametka sobiratelja", observed that singers "do not always sing *byliny* in exactly the same way". Rybnikov attributed this to such reasons as the personal mood of the singer at the moment of performance, learning one *bylina* from several teachers, a bad memory.[9]

In the next decade, the 1870's, A. F. Gil'ferding made an even more significant collection and study of the *byliny*, collecting both in the same region as Rybnikov and going still farther north than his predecessor. Rybnikov's and Gil'ferding's work inspired a systematic search for *bylina* texts in Russia, a search that has lasted up to the present time, in addition to emphasis on the repertory and stylistic characteristics of individual performers. It is primarily Gil'ferding's approach that has persisted with little change for the past one hundred years and his conclusions that most frequently have been restated, reinterpreted or refuted. Gil'ferding maintained that a *bylina* consists of two component parts: typical elements, which are mostly descriptive in content or contain speeches placed in the mouths of heroes; and transitional elements, which connect the typical elements and through which the course of the action is narrated. The first of these elements, according to Gil'ferding, consists of the *loci communes* or petrified verbal formulas, which are transmitted from singer to singer. Gil'ferding believed that the singer knew these passages by heart and sang them in the same way, no matter how many times he repeated the *bylina*. He noted, however, that the typical elements of one singer differed from those of another. Gil'ferding maintained that the transitional elements, as contrasted with the typical elements, probably were not learned by heart, but were composed during performance on the basis of a framework fixed in the singer's memory. He concluded that each singer "selects for himself from the mass of ready epic

[9] P. N. Rybnikov, *Pesni, sobrannye P. N. Rybnikovym*, 2nd ed. (Moscow, 1909), I, xciv-xcvi.

pictures a more or less sizeable stock, depending on the strength of his memory, and having memorized them, uses this stock in the same way in all of his *byliny*". In spite of the fact that each singer assured him that he had sung his *byliny* just as he had heard them, Gil'ferding emphasized the personal element or a singer's individual manner, which is apparent almost in every *bylina*. In reference to two singers who had learned *byliny* from the same teacher, he observed that their songs were "very similar in content" but significantly different "in execution and turns of speech". Gil'ferding also went so far as to point out that a singer changes a *bylina* each time that he sings it by adding, substituting or leaving out certain particles, words, expressions and whole lines.[10]

By the end of the nineteenth and beginning of the twentieth centuries a number of important *byliny* collections had been published and many special studies undertaken.[11] Soviet scholars also continued the traditions of the past, especially in regard to collecting in the field. The first major expedition of the Soviet era, which was organized under the direction of the Sokolov brothers toward the end of the twenties of this century and called "In the Steps of Rybnikov and Gil'ferding", collected materials from the same area as its distinguished predecessors.[12] Since that time there have been numerous expeditions, so that certain areas have been covered again and again and the repertory of a single singer has frequently been written down several times, with a time interval of ten or more years.[13]

Since the time of Rybnikov and Gil'ferding, collectors have continued to show nterest in the singer. In a small volume published in 1924, B. M. Sokolov, like many of his predecessors, occupied himself with the problem of the "special talent" that a singer of oral epic songs must have, observing that this special talent involves an exceptional memory, but adding that memory alone is not sufficient. What is also needed, according to Sokolov, is the ability to learn a new *bylina* quickly. In illustration of this he recounted the collector Ončukov's experience with Pëtr Pozdeev, who gave more *byliny* to Ončukov than any other

[10] A. F. Gil'ferding, *Onežskie byliny*, 4th ed. (Moscow-Leningrad, 1949), I, 56-65.

[11] See especially N. S. Tixonravov and V. F. Miller, *Russkie byliny staroj i novoj zapisi* (Moscow, 1894); A. V. Markov, *Belomorskie byliny* (Moscow, 1901); A. D. Grigor'ev, *Arxangel'skie byliny*, I (Moscow, 1904), II (Prague, 1939), III (St. Petersburg, 1910); N. E. Ončukov, *Pečorskie byliny* (St. Petersburg, 1904); V. F. Miller, *Byliny novoj i novejšej zapisi* (Moscow, 1908). Among the important studies are A. N. Veselovskij, *Istoričeskaja poètika* (Leningrad, 1940); V. F. Miller, *Èkskursy v oblast' russkogo narodnogo èposa* (Moscow, 1892); A. P. Skaftymov, *Poètika i genezis bylin* (Saratov, 1924).

[12] Ju. M. Sokolov, "Po sledam Rybnikova i Gil'ferdinga", *Xudožestvennyj fol'klor*, II-III (1927), 3-33.

[13] A part of this material has been published in the following large collections: A. M. Astaxova, *Byliny Severa*, 2 vols. (Moscow-Leningrad, 1938-1951); Ju. M. Sokolov, *Onežskie byliny* (Moscow, 1948); È. Borodina and R. Lipec, ed., *Byliny M. S. Krjukovoj*, 2 vols. (Moscow, 1939-1941); and others.

singer. Pozdeev was so interested in this collector's work that he was present even when other singers sang. On one occasion, according to Ončukov, the singer Durkin sang a *bylina* about Luka and the dragon in the presence of Pëtr Pozdeev. When Pozdeev and Ončukov met several days later, Pozdeev asked if the collector would like to listen to him sing a new *bylina*, about Luka and the dragon. Ončukov reported that Pozdeev remembered the *bylina*, after having listened to it only once. Ončukov concluded by remarking that Pozdeev was 65 years old. By way of further illustration Sokolov added the experience of Gil'ferding with Matrëna Men'šikova, who sang for him from beginning to end a Serbian song about Iova and Mara. Sokolov remarked that what made this performance so amazing was that although the singer was illiterate, she nevertheless learned this song, which she probably heard read by her husband or son from a book in the translation of the poet Ščerbina. Sokolov added to this another observation made by Gil'ferding regarding how quickly a gifted singer can learn a *bylina*. He reported that Nigozerkin, a 40-year-old singer from the Olonec province, learned a *bylina* he hadn't heard before from an old man from Kenozero who dropped by to spend the night.[14] After citing the above examples, Sokolov pointed out that the reason a singer can remember a number of verses unusually large, even for an exceptional memory, was due to the fact that he has at his disposal a stock of ready-made descriptions, pictures, etc., which he transfers from one *bylina* to another or repeats several times in the same *bylina*. He added that these are the reasons that make learning and remembering a long *bylina* easy and that also make clear the very method of learning a *bylina*. Sokolov concluded these observations by maintaining that the singer does not memorize a new *bylina* but remembers its basic content, and when he begins to sing it he uses the descriptions, fixed comparisons and turns of speech already familiar to him.[15]

Worthy of special mention is the research carried out by the Soviet scholar A. M. Astaxova.[16] Astaxova has analyzed the *bylina* tradition of the last one hundred and fifty years, studying in particular the creative process of the folk epic, along with the significance of the environment and the influence of written literature. Astaxova added nothing new to our concept of how songs in the Russian oral epic tradition are composed, however, when she attributed variants or "contaminations" of plots and episodes to an expression of the creative attitude of the singer to his material, to the influence of certain social tendencies, or simply to the fact that the singer is a bad one and forgets.[17] Astaxova did,

[14] B. M. Sokolov, *Skaziteli* (Moscow, 1924), pp. 42-46.
[15] *Ibid.*, pp. 68-69.
[16] See especially *Russkij bylinnyj èpos na Severe* (Petrozavodsk, 1948).
[17] *Byliny Severa*, (Moscow-Leningrad, 1938), I, 33.

however, take us a step further in our concept of the lack of stability of a *bylina* when she classified the singers of the north of the nineteenth and twentieth centuries into three categories, based on how they interpret and recreate *byliny*. In the first category she included those singers who reproduce songs exactly or almost as transmitted, admitting, however, that even such songs exhibit certain individual features. Astaxova labeled the singers in this group as basically only performers who transmit with more or less talent the works learned by them, even though certain of them are distinguished by great artistry of performance. Astaxova concluded that when this type of singer begins to forget, he usually shortens the *bylina* but doesn't change those parts that are left fixed in his memory. The second category of singers has been defined more or less in Gil'ferding's terms, that is, the singer learns primarily only the general outline and by selecting typical elements and freely creating transitional elements he works out his own fixed text, molding it into a definite composition and verbal form, which remains basically unchanged and varies only in details. In this way, new wordings of *bylina* plots arise. A singer in this group contaminates the plot of one *bylina* with that of another when he forgets. Astaxova described the third type of singer as a kind of improviser who neither creates nor works out a fixed text, but who changes the text each time he sings, using the entire arsenal of plots, episodes, motifs, images, formulas. Each text will significantly deviate from an earlier given variant on the same subject, sometimes with complete reconstruction of the entire composition, with contaminated places, with changes in the names of heroes, etc. This type of singer, according to Astaxova, introduces completely new descriptions or episodes and reconstructs the plot: the denouement is changed, plots are contaminated, scraps of forgotten *byliny* are attached to personages with whom they do not belong.[18] Astaxova leaves us with an even greater feeling for the lack of stability of the *bylina*; however, we are still left to think in terms of a singer willfully changing things or unwillfully forgetting.

In recent years, P. D. Uxov has devoted special attention to typical elements as a key to identifying *byliny* of unknown authorship. Uxov reaffirms Gil'ferding's observations by stating that even though the content of typical elements might coincide in general features, each singer's verbal expression of it is different from that of all other singers. Uxov concludes that these elements are not learned by heart because of the lack of correspondence between typical elements from singer to singer. He maintains that in learning these parts of a song the singer replaces everything unclear, not comprehensible to him, with things familiar. In this way he explains the presence of Cossacks from the Don, peasants, etc. at Vladimir's feast. Through frequent repetition of *byliny*, singers

[18] *Ibid.*, pp. 71-82.

usually work out a fixed text of typical elements, which pass with almost literal exactness from one plot to the next. Regarding those singers who improvise, Uxov remarks that they work out a fixed text for the majority of typical elements, but because certain of them are not fixed in the memory, their texts can be unstable. Uxov also adds that the typical elements of a singer, in spite of their stability, are not completely unchangeable.[19] Through concentration primarily on typical elements or *loci communes*, Uxov has emphasized the fact that even these are not learned by heart,[20] and he has also suggested in connection with this the composition of these parts of a song by analogy, limited though it may be.

Those scholars above to whom special attention has been given represent an approach that began in the second half of the nineteenth century and that has continued to the present. It has been known for a long time now that the content of a *bylina* from performance to performance is unstable, and therefore not memorized. In recent years it has been pointed out that those elements of oral epic song that have been termed *loci communes*, typical elements, petrified verbal formulas, etc., are not memorized. These issues have been resolved. But when scholars have been faced with the problem of how an oral poet actually composes his songs, agreeing now on the point that these songs are not memorized word for word, they have continued to conclude that a singer willfully chooses, shifts or manipulates one element or another, does something to a song for reasons that depend on the singer himself, on his good or bad memory, on his desire to change a song for one reason or another.[21] It is these conclusions, which are based on the concept of a fixed text or a fixed group of texts to which a poet does something, that will be reexamined. By applying a different system of analysis, by looking at the *bylina* from a different point of view, certain findings of studies previously made no doubt will be corroborated, while others,

[19] P. D. Uxov, "Tipičeskie mesta (*loci communes*) kak sredstvo pasportizacii bylin", *Russkij fol'klor*, II (1957), 130-152. See also his article "Iz nabljudenij nad stilem sbornika Kirši Danilova", *Russkij fol'klor*, I (1956), 97-115.

[20] See Wildrid Chettéoui, *Un Rapsode Russe; Rjabinin Le Père: La Byline au XIXe Siècle* (Paris, 1942), pp. 68-69: "De plus, nous ne pensons pas que le diseur apprenne par cœur les pièces mobiles et nous verrons que, contrairement à l'opinion généralement admise, elles sont sujettes à des variations internes et subissent une évolution."

[21] Attention should also be called to H. M. and N. K. Chadwick, *The Growth of Literature* (Cambridge, 1936), II. This work treats certain aspects of the practice of oral epic poetry, but, as Professor Lord has pointed out, Chadwick failed to "see the *necessary* connection between the practice which he describes and the form of the poetry". A. B. Lord, "The Singer of Tales", (unpubl. diss. Harvard, 1949), Chapter I. The following statement made by the Chadwicks is worthy of note: "It would seem, then, that what one singer transmits to another is in the first place the *bylina* technique, and in the second, the sphere of action within which the hero's activity is confined" (p. 248). See also C. M. Bowra, *Heroic Poetry* (London, 1952), especially chapters V-XII.

it is hoped, will be supplemented or redefined and made more exact. A basic task will be to see just how Parry's concept of the formula fits the Russian line of oral epic verse, and to redefine the formula if necessary, giving those features distinctive to this tradition. Certain peculiarities of formulas will be considered, along with the role of parallelism. This kind of examination, shifting focus at times from the individual singer to the surrounding tradition, should help us to discover how a singer in the Russian oral epic tradition composes his lines of verse.

III

THE FORMULA

Milman Parry has defined the "formula" as "a group of words which is regularly employed under the same metrical conditions to express a given essential idea".[22] This study will try to determine the formulaic features of Russian oral narrative poetry and attempt to demonstrate the usefulness of this method of composition in the Russian tradition of oral epic. One of the problems to be faced will be the distinction between conscious memorization and repetitions which come about from solving the same problem over and over again. This tradition, therefore, will be examined both from the point of view of one singer, concentrating specifically on the songs of Trofim Grigor'evič Rjabinin, and also from the standpoint of a specific moment of performance, using for this purpose the songs of Rjabinin which were collected by A. F. Gil'ferding in the summer of 1871 in northern European Russia, in the Onega region.[23] By focusing attention on the songs of an individual singer, it should be possible to observe the formula at work, in addition to seeing part of a single singer's repertory of formulas, which does not necessarily correspond to the formula repertory of other singers. In this regard, it has been repeatedly suggested that the individual singer is the formula maker, the establisher of his own formula habits, and that although his formulas might have existed at one time or another within the history of the tradition, these formulas become operative for him only when their regular use in his songs is established, their number varying with an increase or decrease in his thematic repertory.[24]

Before proceding to the texts and analysis of formulas, it is necessary to

[22] Parry, *op. cit.*, p. 80.

[23] A. F. Gil'ferding, *Onežskie byliny*, 4th ed., (Moscow-Leningrad, 1950), II, 1-166. Throughout this study the texts form this volume, including those of other singers as well as those of Rjabinin, will bear first the song number followed by the lines referenced, i.e., 80:736-739. Upon occasion, for the sake of clarity, the song number will be preceded by the letter "G", designating the Gil'ferding as opposed to another version, i.e., G 80:736-739.

[24] Lord, *op. cit.*, pp. 30ff.

describe briefly the metrical characteristics of the Russian *bylina*, since the length and internal composition of the formula will depend in part on these features.[25] In an article entitled "Studies in Comparative Slavic Metrics", Roman Jakobson already has analyzed in great detail the meter, and only those points relating to the structure of the formula will be mentioned here.[26] Russian epic verse shows considerable syllabic oscillation. The longer line, which is more usual in the *bylina*, ranges, as a rule, from ten to sixteen syllables, with alternation of accented and unaccented syllables. The line is organized by three primary accents, of which the first and the third are the strongest. The external syllables of a line, those before the first and after the third primary accent, cannot carry strong stress. The number of unaccented syllables between accented ones varies in general from one to three. The line has no fixed internal break. The rhythmical pattern can be summarized as a line with a fluctuating number of syllables,[27] three primary accents,[28] and a moveable break.

There is close interplay between the melody and the meter.[29] Compensation is made in regard to both of these restrictive elements. The melodic line can be adjusted to the line of verse, for instance, by increasing or decreasing the length of time a syllable is held. On the other hand, the verse line can be lengthened or shortened syllabically to fit the melodic line. A singer's formulas seem to submit readily to expansion or contraction, especially through the use or omission of meaningless particles, pronoun fillers and repeated prepositions,[30] in addition to the use of expanded or contracted morphological endings. We are forced to rely on scant and less than precise information concerning the actual melodies

[25] Those parts of the Russian song texts quoted in the body of this study have been transliterated with all their phonetic peculiarities. Accents have been preserved if they appeared in the texts used. In certain rare instances, for the sake of convenience, I have normalized such forms as *oni, ony, eni, ëny* when this form was the only distinctive one in several examples.

[26] Roman Jakobson, "Studies in Comparative Slavic Metrics", *Oxford Slavonic Papers*, III (1952), 35ff. For a thorough and well-illustrated discussion of patterns of accent distribution within lines of Russian epic verse with fluctuating syllable count see especially pp. 39-48 of this article. See also Jakobson's article, "The Kernel of Comparative Slavic Literature", in *Harvard Slavic Studies*, I (1953), 24-31. For a discussion of existing theories of popular versification consult M. P. Štokmar, *Issledovanija v oblasti russkogo narodnogo stixosloženija* (Moscow, 1952).

[27] The fact that Russian songs are no longer sung to the accompaniment of a musical instrument probably leads in part to greater freedom in the number of syllables in a line.

[28] See V. M. Žirmunskij, *Vvedenie v metriku* (Leningrad, 1925), p. 243. Žirmunskij, among others, gives a slightly different description of the line. He states that the line of the *bylina* has four main metrical accents, two per hemistich. In a footnote on p. 283, Žirmunskij adds that those scholars who do not accept the dactylic ending with two accents admit three rather than four main accents per line.

[29] See T. V. Popova, *Russkoe narodnoe muzykal'noe tvorčestvo* (Moscow, 1955). Reference is to the first edition only of this work.

[30] This view of the function of repeated prepositions is contradicted by A. P. Evgen'eva in *Očerki po jazyku russkoj poèzii v zapisjax XVII-XX vv.* (Moscow-Leningrad, 1963).

of those *byliny* that will be subjected to analysis. Rybnikov commented that Trofim Grigor'evič Rjabinin sang his *byliny* to two melodies.[31] Gil'ferding's *Onežskie byliny* can be consulted for a musical transcription of the beginning of two songs by Rjabinin. The two songs transcribed are "Volga and Mikula" and "Dobrynja and Vasilij Kazimirov".[32]

Most lines can be further described in terms of initial and final parts, with a 2:1 accent distribution. These parts can in turn be described from the standpoint of word groupings and their syllable count. Prepositions, particles and conjunctions are either enclitic or proclitic and should not be counted as separate words but as syllables of the preceding or following words. The internal rhythm of the formula is indicated by syllabic or word groupings around from one to three primary accents, depending on the length of the formula. In this study, a whole line of verse is considered to be the maximum length of a formula. Unlike Yugoslav oral epic poetry, the line is marked by a moveable break. The position of the moveable break in the line is determined by the boundary between the initial and final formula. The length of the formula then coincides either with the moveable break or with the end of the line, which is marked by a syntactic pause and in most cases also by a dactylic ending. The rhythmical structure of the melody and meter and the normal accents of the speech material, which usually coincide, complete the description of the metrical background of formula structure.[33]

Certain syntactic patterns recur so frequently that they also may be classed, along with metrical conditions, as a basic restrictive element in formula construction. An examination of certain specific syntactic patterns produces some interesting results. The first 1,000 lines of song No. 80 by Rjabinin were scanned for whole-line prepositional phrases. Under this heading I have considered only those lines that contain a single preposition, even though it might be repeated several times in the line, and its object, which can include one or more modifiers and one or more noun-objects: for example, *I ot togo ot knjazja ot Vladymira* (80:71), "from prince Vladimir". Particles, repeated prepositions and pronoun "fillers" have been exlcuded from the syntactic description. The 1,000 lines scanned contain ninety-eight whole-line prepositional phrases, approximately ten per cent. These ninety-eight lines fall into sixteen syntactic patterns. One of the more frequently recurring patterns is "preposition-modifier-object-

[31] Rybnikov, *op. cit.*, p. xciii.

[32] Gil'ferding, *op. cit.*, II, insert between pp. 80 and 81. For comment on this transcription see pp. 784-785.

[33] Žirmunskij, *op. cit.*, p. 243: "Russian folk verse is sung verse, that is, its rhythmical character is determined first of all by the rhythmical structure of the melody; however, at the same time the musical accents are distributed according to word (or phrase) accents, not violating the speech material but making great use of the accentual doublets that exist in folk dialects...."

object", which accounts for about twenty-three percent of these ninety-eight lines: *Ko tomu ko knjazju ko Vladimiru* (80:775), "to prince Vladimir". The modifier is either a descriptive adjective or a demonstrative pronoun. In this pattern both objects form an appositional group and refer to the same person or thing. The three appositional groups within this pattern are common noun and proper name, e.g., prince Vladimir; proper name and patronymic or last name, e.g., Alëša Popovič; common noun referring to a general category plus common noun designating a specific thing within this category, e.g., the way the road. In only one instance is there an example of proper name and common noun: see *Da j vo slavna Svatogora vo bogátyrja* (80:192), [with the strength] "of the famous hero Svjatogor". This pattern can be explained by the accent of the final word in the line, which frequently is shifted back to the second syllable by Rjabinin, making this word suitable for final position in the line, at the same time preserving the dactylic ending. In the texts of Rjabinin at our disposal there is no occurrence of an accent shift in *Svjatogor*, therefore, this word would seem to be unsuited for final position in the line, thus forcing this exceptional syntactic pattern. The second most frequently recurring pattern is "preposition-modifier-object-modifier", which accounts for about thirty-nine per cent of these ninety-eight lines: *Da za ètyi za gory za vysokii* (80:255), "behind the high mountains". The first modifier in this pattern is a demonstrative or possessive pronoun, and rarely a descriptive adjective. The second modifier, the last element of the line, is always a descriptive adjective. Thus, about sixty-two per cent of these whole-line prepositional phrases fall into two basic patterns. The remaining thirty-eight percent are distributed among fourteen different patterns, which occur from one to six times. Statistics, of course, will vary from one sampling to the next, but they can help to indicate that there are recurrent patterns which should be taken into account.

If the remarks accompanying Chart I are consulted, especially those to line 249, it can be seen that certain of the formulas which a specific singer uses seem to have a fixed position in the line. However, if additional lines of this same singer are examined, it becomes obvious that some other formulas are not limited in position to the initial or final place in the line. The position of certain formulas may be interchangeable, sometimes appearing in initial position, and at other times in medial or final position. This is especially true of proper names. Compare the positions of *Il'ja Muromec* in *Il'ja Muromec da syn Ivanovič* (74:161), "Il'ja Muromec, son of Ivanov", *Prixodit Il'ja Muromec vo bel šatër* (75:237), "Il'ja Muromec arrives [and enters] the white tent", and *Aj že staryja kazak da Il'ja Muromec* (74:163), "Oh, old Cossack, Il'ja Muromec"; and also *Vladymir knjaz'* in *Vladymir knjaz' po gorenki poxažival* (81:8), "Prince Vladimir paced the floor of the chamber", *Govoril-to ved' Vladymir knjaz' da takovy slova*

(74:209), "Prince Vladimir said as follows", and *Govoril-to emu Vladymir knjaz'* (80:831), "Prince Vladimir said to him". In the analysis that follows, however, only initial, final and whole-line formulas will be considered, assuming that these are the units most useful to the singer in the composition of his songs.

The four elements that combine to make a formula can be summed up as follows: first an essential idea to express; second, the space into which the idea must be fitted, its length; third, the internal rhythmic pattern or metrical content, which includes syllabic or word groupings around primary accents; and fourth, the syntactic pattern. In describing any given formula these four elements should be considered. Thus, the formula *Govoril Vladymir takovy slova* (line 249 in Chart I) would be described as an eleven-syllable formula with a metrical content of three-three-three-two and three primary accents, and with a syntactic content of verb-subject-adjective-direct object, expressing the idea "Vladimir said as follows". The whole-line formula is in turn composed of two formulas: one is an initial six-syllable formula with a metrical content of three-three and two primary accents, and with a syntactic content of verb-subject, expressing the idea "Vladimir said"; and the other is a final five-syllable formula with a metrical content of three-two and one primary accent, and with a syntactic content of adjective-direct object, expressing the idea "as follows".

The theory behind the concept of composition by the use of formulas or formulaic expressions is simply that any idea required by a particular tradition finds expression in phrases metrically useful to a singer of that tradition. The important feature of the formula, thus, is not the fact of its repetition but the nature of its function as a metrical component. To be useful to a singer as he composes rapidly, a phrase or word must suggest to him that it belongs at only a certain point in this verse line, some phrases or words being useful to the singer at more than one point in the line, others at only one point; that is, it must be a significant segment of the rhythm. To be useful to the singer, every phrase or word which is metrically significant should also be a syntactic entity; that is, if it is not a polysyllable which by itself makes a half-line verse, it should at least be a phrasal group or clause. The usefulness of any given expression, thus, is in direct proportion to the number of times a singer uses it. In the Russian tradition, we can expect less possibility for exact repetition because of the loose verse structure.

Formula analysis of a specific text will serve as illustration of what has already been said, and will also indicate how extensive the formulas are in a given sample. These results should be accepted as final only with caution because of possible normalization of oral texts, because of inaccuracies due to non-scientific methods of recording these texts, and because of certain subjective judgments on my part. When possible, I have indicated those passages in the texts that are known to be mechanically repeated on the part of the recording scribe and that do not necessarily represent an actual word-for-word repetition on the part of the singer. In the notes to Gil'ferding's *Onežskie byliny* these places are indicated on the basis of marginal notes in the manuscripts. Nevertheless, these mechanical repetitions will not significantly change the picture. For analysis I have chosen fifteen lines from song No. 79 by Trofim Grigor'evič Rjabinin, an unlettered singer from the Kiži district in the Onega region who was born in Garnicy and who later resettled in Seredka. The passage selected is from the middle of a song about Dobrynja and the dragon. This song consists of 470 lines and was recorded by Gil'ferding in St. Petersburg on the 24th of November, 1871. Rjabinin was 78 years old at the time of recording, and an accomplished performer with a large repertory. The analysis will use as supporting evidence a total of 5,420 lines which comprise the eighteen *byliny* of Rjabinin that Gil'ferding collected and published in *Onežskie byliny*. Material included in the appendix to this volume will not be used. About ten years prior to 1871, P. N. Rybnikov collected songs from this same singer.[34] These texts will be used only for occasional reference. Supporting evidence for the particular type of underlining in Chart I, including formula types and their prevalence, is presented in the appendix at the end of this study. A solid line in the chart indicates a formula, that is, the repetition is word-for-word the same. In this case, simply a reference to the text and line in which the phrase is repeated is given. A broken line indicates a formulaic expression. In addition to reference to text and line, examples of the systems or patterns to which this type of phrase belongs are given. I have also included in the supporting evidence examples involving some similarity of phraseology, but not enough to warrant underlining on the chart. A formulaic expression must share with other expressions the same basic patterns of rhythm and syntax and must also have in common with these other expressions at least one word in the same position in the line.[35] The formula and formulaic underlinings in this study relate only to a specific number of lines of one singer. If the songs recorded by Rybnikov or

[34] Rybnikov, *op. cit.*, pp. 3-141. Throughout this study the texts from this collection will be designated first by the letter "R", followed by the song number and the lines referenced, i.e., R 8:1-5.

[35] A concept similar to the definition of formulaic as used in this study, but more limited, is expressed by Robert Austerlitz' use of the term "parallel" to indicate repetition that is not word-for-word exact in two lines in the immediate vicinity of one another, i.e., in consecutive lines: "A line is said to be parallel to another line when the words which occur within it are identical with and in the same position as the words of the other lines, with the exception of one or two words (in each of the two lines). The one or two words which are not identical must nevertheless occupy the same position within the respective lines." *Ob-Ugric Metrics: The Metrical Structure of Vogul Folk-Poetry* (= *Folklore Fellows Communication*, No. 174, Helsinki, 1958), p. 45.

CHART I

Govoril Vladymir|takovy slova:	1; 2, 3	Vladimir said as follows:
Aj ty staryja kazak|da Il'ja Muromec!	4; 5, 6	"Old Cossack, Il'ja Muromec!
A ty s"ezdi-tko vo zemlju-tu|vo švedskuju,	7; 8, 9	Go to the Swedish land,
A ko tomu korolju ty s"ezdi|k švedskomu,	10; 11, 12	Go to the Swedish king,
Otvezi-tko dani|za dvenadcat' god,	13; 14, 15	Collect tribute due for twelve years,
Za dvenadcat' god|da s polovinoju.	16; 17, 18	For twelve and a half years."
Tut Olešen'ka Grigor'evič|po gorenke poxaživat,	19; 20, 21	Olešen'ka Grigor'evič paces the chamber,
Poslovečno knjazju|vygovarivat:	22; 23, 24	Speaks out to the prince:
Ty Vladymir knjaz'|da stolnë-kievskoj!	25; 26, 27	"Prince Vladimir of royal Kiev!
A j nakin'-ko tu ved' slúžobku|velikuju,	28; 29, 30	Bestow the great task,
Da veliku slúžobku|nemaluju,	31; 32, 33	The great task which is not an insignificant one,
Na togo da na molódago|Dobrynjušku,	34; 35, 36	On young Dobrynjuška,
Čtoby s"ezdil on v daleče|čisto pole,	37; 38, 39	Of going far off into the open plain,
Ko tym slavnym ko goram|da soročinskiim	40; 41, 42	To the famous Soročinsk mountains,
Da sxodil by on vo nory|vo zmeinyi,	43; 44, 45	Of going into the dragons lair ..."

those variants and fragments of variants appended to the Gil'ferding collection had been included in this analysis, or if the songs of other singers had been included in addition, there is little doubt that the evidence for formula and formulaic expressions would be significantly increased, and that some of the expressions which have received formulaic underlinings would receive the solid line indicating clearly a formula.

The whole- and half-line phrases that occur more than once in 5,402 lines of verse from a given singer have been underlined in Chart I. This chart shows that in relation to 5,402 lines from one singer a certain number of phrases in a given passage is formula. Although these statistics are relative, and would most likely vary were this kind of analysis to be applied to other passages, they nevertheless will help indicate some of the underlying patterns that a singer uses to express his ideas. Of the fifteen whole lines, 33.3 per cent are formula, 60 per cent are formulaic, and the one remaining unmarked line was found to be neither formula nor formulaic, on the basis of the material used. In relation to half-line formulas, 66.6 per cent of the fifteen lines contain half-line formulas, and 26.7 per cent contain half-line formulaic phrases. In other words, about 55.6 per cent of the material analyzed from a given singer on the basis of 5,402 lines is formula, and about 37.8 per cent is formulaic. To be noted is the fact that there was only one whole line and parts of two lines that did not fit into either a formula or formulaic pattern, and I say this with reservations. In certain instances the supporting evidence was not abundant, but there was still enough evidence, in my opinion, to justify labeling these phrases as formulaic. The evidence for formula and formulaic expressions would most likely increase if more texts of this singer had been available to work with, and this increase would be even more of a probability if the songs of other singers had been used, especially of those singers in the same or neighboring districts, as is indicated by certain comments below on the supporting evidence. It can be concluded, based on the results of this analysis, that the half-line unit is more useful to the singer than the whole-line unit in the composition of his songs. The results also lead us to conclude that the Russian bards recompose rather than memorize, based on the high percentage of formula and formulaic expressions found in this passage.

The 5,402 lines used in this analysis constitute eighteen different songs, ranging from 57 to 1,022 lines, one of the longest song texts obtained by Gil'ferding. Rybnikov's text of the same song by Rjabinin, actually the longest song text obtained by this collector, consists of 926 lines. These songs collected by Gil'ferding were all recorded from musical recitation. Gil'ferding's method of obtaining the texts was to have the singer sing his songs at a slower tempo and with more pause than is usual at the end of each line in order to facilitate the process of taking down each line as accurately as possible.[36] The corpus of Rjabinin's song texts available to us in Rybnikov's collection has not been used as supporting evidence for formula and formulaic phrases, although some of this material will be used from time to time for comparison. Rybnikov's texts represent songs that were first recited and then corrected on the basis of one or several singings, which took place sometimes immediately afterwards and sometimes on different occasions.[37] Gil'ferding's texts come closer to representing an accurately fixed text based on one specific moment of performance, rather than a text normalized on the basis of several performances, both recited and sung. A more successful fixing of a text during a specific moment of performance, without the aid of mechanical recording devices, was to be achieved later by the Sokolov expedition, which collected songs from the same area as Gil'ferding and Rybnikov in the twenties of this century.[38] This attempt to record the texts shed new light on the problem of accurately transcribing a text, in addition to giving new evidence that a singer does not necessarily sing his lines in exactly the same way when he repeats the performance of a song.

Some additional remarks on Chart I might make certain judgments clearer, and, in certain instances, present the possibility of slightly changing the results. Line 249 belongs to a type of line that occurs frequently in oral narrative poetry, one that introduces speeches.[39] In Rjabinin's songs the syntactic pattern for this whole-line expression is almost consistently verb-subject-adjective-direct object. In only four instances does some form of the verb *govorit'*, "to say", appear other than as the first but never the final element of a line, discounting particles, in the more than 200 lines containing this verb. One such instance

[36] Gil'ferding, *op. cit.*, I, 64-65.

[37] Rybnikov, *op. cit.*, pp. xlvi and lxxx-lxxxi.

[38] Ju. M. Sokolov, "Po sledam Rybnikova i Gil'ferdinga", *Xudožestvennyj fol'klor*, II-III (1927), 11-12. Recording of the texts was made with a battery of scribes. Two scribes wrote down the first half of the line, going a little past the caesura, and two more wrote down the second half of the line, beginning a little before the caesura. Later, a third scribe was added to each group. When it was observed that certain defects (i.e., omissions) appeared in the middle of the line under this system, another method was adopted. According to the new method, one scribe took down the beginning of the line, one the middle, and one the end. Sokolov remarked that the line as a whole was not recorded because this would have involved having the singer either stop at the end of each line or force him to sing much slower than usual. He expressed the opinion that experience had not yet taught us that an accurate and good text could be obtained from a singer only when the conditions the singer was accustomed to when singing were preserved. He further commented that in this collective undertaking one could see how subjective the perception of the individual collector sometimes was. There often arose heated arguments in relation to various sounds, words, and particles in particular, and a repeated performance on the part of the singer for the purpose of verification of the text was not always reliable, since these particles varied from performance to performance.

[39] See M. O. Gabel', "Forma dialoga v byline", in *Naukovyj Zbirnyk* (= Naukovi zapysky naukovo-doslidočnoj katedry istoriji ukrajins'koj kul'tury, No. 6, Xar'kiv, 1927), 315-328.

is in the following line: *Tut sobaka Kalin car' govoril Il'je da takovy slova*, "The infidel, czar Kalin, said as follows to Il'ja".[40] This line is irregular in appearance, but is, most likely, a combination of a perfectly regular line, such as *Govoril Il'je da takovy slova*, "He said as follows to Il'ja", with part of a defective *line*, *tut sobaka Kalin car'*, "the infidel, czar Kalin".[41] Parry expressed one of the peculiar situations of oral narrative poetry when he remarked that a singer in an oral tradition who must rapidly compose his songs without stopping will often make irregular lines, not because of himself, but rather because of his technique of composition.[42] It is worth noting that all examples of the verbs *vygovarivat'* and *pogovarivat'*, both close in meaning to *govorit'*, appear in their various forms as the final element in the line. The accents of these verbs, on the third syllable from the end in the infinitive form, make them especially suitable for a position at the end of a line. In each case this line is coupled with a preceding line in which the final element is some form of the verb *poxaživat'*, "to pace". In all but one case the latter verbal form is parallel to the verbal form at the and of the following line. Compare lines 255 and 256 in the chart. These are all examples of formulaic couplets, the first line of which expresses the idea "someone paces, paced, or started to pace in his chamber", and the second line of which expresses the idea "he says, said, or started to say (to someone)". This formulaic couplet serves a purpose similar to the type of line under discussion, that is, it introduces a speech. It is, however, less frequently used by Rjabinin. On the other hand, all examples of the verbs *vosprogovorit'*, *vozgovorit'* and *isprogovorit'*, also similar in meaning to the verb *govorit'*, appear in their various forms as the first element in the line. The accents of these verbs, in the infinitive form on the final syllable, would exclude them normally from a position at the end of the line. These lines, similar to line 249, all end with the expression *takovy slova*, "as follows". *Takovy slova* appears as a final formula in phrases that begin with one of the several verbs with the meaning "to say", and whose position is in the initial part of the line. The only occurrence of this formula at the beginning of a line is in the following line: *Takovy slova Ščurilo pogovarivat*, "Ščurilo says as follows". This line ends with a form of the verb *pogovarivat'*, which consistently appears at the end of the line. Because of the singer's use of *poxaživat* in the preceding line, he should then be expected to use some form of the verbs *pogovarivat'* or *vygovarivat'* at the end of the next line, which he does. The force of parallelism in line structure seems to have been a decisive one. Before leaving

[40] See the appendix notes to the appropriate lines in the chart for those line references that do not appear on this page and those that follow.

[41] For a discussion of this kind of line, a complete line plus part of another, see F. N. Korš, *O russkom narodnom stixosloženii* (= *Sbornik otdelenija russkago jazyka i slovesnosti*, LXVII, St. Petersburg, 1901), pp. 28ff.

[42] Parry, *op. cit.*, p. 140.

line 249, we might consider the usefulness of this type of system to a singer. The verb and direct object remain constant at either end of the line, and into the three-syllable space between can be fitted any number of subjects:

	Vladimir	
	Dunaj da	
	Il'ja da	
govoril/govorit		takovy slova
	Ivan da	
	korol' on	
	Samson da	

The second line of the chart, 250, is a whole-line vocative expression. Syntactically, it is basically the same as a whole-line phrase that names a person and consists of an adjective plus a common noun describing the general category to which the person being named belongs plus his or her proper name and patronymic. This description does not take into account particles and pronoun fillers. Thus we have "the young wife Nastas'ja Mikulična", "the honorable widow Mamel'fa Timofeevna", "the czar's son Fedor Ivanovič", etc.[43]

The initial part of line 251 could be classified as formula rather than formulaic, since the only difference in the comparisons involves initial particles and the position of the word boundary between the initial and final parts. The initial particles, *a* versus *aj*, are virtually the same, both being monosyllabic. Compare the word boundaries toward the ends of these lines:

No. of syllables	
8 (4 plus 4)	vo zemlju-tu // vo švedskuju
8 (3 plus 5)	vo zemlju // v politovskuju

The basic syntactic pattern of this line can be further described in general terms as a whole-line expression which contains a verb (whose subject is contained in a whole-line expression immediately preceding this line) plus a complement consisting of a prepositional phrase (preposition-object-adjective). Line 263 in the chart represents a similar pattern.

[43] Cf. Ego bol'šij syn Ivanuško Ivanovič (90a:11), Da čestna vdova Mamel'fa Timofeevna (80:200), Ty čestna vdova Mamel'fa Timofeevna (80:735, 762, 792), To čestna vdova Nastas'ja da Vasil'evna (85:525, 535), To čestna vdova Nastas'ja-to Vasil'evna (87:291), Ty čestna vdova Nastas'ja da Vasil'evna (85:242, 531), A 'šče groznyj car' Ivan Vasil'evič (88:23), Tut i groznyj car' Ivan Vasil'evič (90a:17), Krestnyj ty moj batjuška Samson Samojlovič (75:260), Molodoj bojarin Djuk Stepanovič (85:13, 27, 33, 64, 73, 96, 108, 135, 166, 202, 210, 222, 245, 265, 296, 319, 336, 341, 347, 373, 384, 397, 410, 412, 421, 452, 478), Tut molodoj bojarin Djuk Stepanovič (85:550), A j tvoja-to moloda žena Nastas'ja Mikulična (80:688), Tut moskovskoj knjaz' Mikituška Romanovič (88:31), Aj že carskoju cyn Fedor Ivanovič (83:100), Carskoj syn Fedor Ivanovič (83:108).

If our own rules are again relaxed, line 252 might possibly be classified as formulaic, based on a comparison with line 251, which immediately precedes it, and similar patterns that can be found in the notes in the appendix. In line 252 there are thirteen syllables and a basic syntactic pattern of preposition-object-verb-adjective. The preceding line contains both the same number of syllables as well as the same syntactic elements; however, the verb is in the first position in the line, the other elements following in the same order, i.e., verb-preposition-object-adjective. If the verb in line 252 is shifted to first position, then there is good formulaic evidence, since the same basic syntactic pattern obtains. Thus, the first half of the line would also become formulaic. If emphasis is shifted from the whole line to part of the line it might then be suggested that *ty s"ezdi k švedskomu* is a telescoping of the preceding line 251, *A ty s"ezdi-tko vo zemlju-tu vo švedskuju*, with omission of the word *zemlju* and use of a different preposition, which is a repetition of the preposition appearing in the initial part of the same line.

Line 253 is a whole-line eleven-syllable formula, consisting of a verb plus direct object plus prepositional phrase. It is very closely related to the eleven-syllable line 254, also a formula, which immediately follows and which is an echo in part of the second half of the preceding line:

Otvezi-tko dani // za dvenadcat' god,
Za dvenadcat' god // da s polovinoju.

The relationship between these two lines can be expressed thus:

A	B	
	B	C

Robert Austerlitz designates this type of structure a "terrace."[44] These two lines comprise a formula couplet. Neither of these formulas, however, is bound to to the other: both can be used independently.

A similar type of relationship can be established between lines 251 and 252:

	A ty s"ezdi-tko //	vo zemlju-tu //	vo švedskuju,
A ko tomu korolju //	ty s"ezdi		// k švedskomu,

This relationship can be designated as follows:

	A	B_1	C
B_2	A		C

[44] Austerlitz, *op. cit.*, p. 65. "A *terrace* is a complex of two lines in which the latter portion of the first line is identical with the former portion of the following line."

Compare also lines 258 and 259:

A j nakin'-ko tu // ved' služobku // velikuju,
Da veliku // služobku // nemaluju,

which can be stated as:

A B C
C B D

Such cases are not considered to be formulaic.[45]

Line 255 received a broken underlining, indicating that it is part of a whole-line formulaic system whose basic pattern is proper name plus patronymic plus preposition plus object plus verb.[46] It is tempting to consider the first part of this line as formulaic, since there is both rhythmic and syntactic agreement between the expressions *tut Olešen'ka Grigor'evič* and *a j Ščuriluško-tot Plenkovič*; however, there is no word in common in the same position in the line. The final part of the line, however, was found to be formula. The usefulness of a verb such as *poxaživat'* at the end of the line has already been discussed in connection with line 249. This line stands in couplet relationship with the line immediately following. These lines are bound to one another by the final verbs, which have the same accent pattern and the same *-iva-* verbal suffix.

Formula evidence has been established for line 256 in terms of both half lines, but not for the whole line, which belongs to a formulaic system, although supporting evidence is scant. This pattern consists of adverb-indirect object-verb. Both in Gil'ferding's and Rybnikov's texts the initial word, *poslovečno*, is attested only in conjunction with some form of the verb *vygovarivat'*, which always is the final element in the line, This verb, however, does appear without *poslovečno*. *Poslovečno* is always in initial position in the line, even in the Rybnikov texts. Compare the following minor exceptions: *I poslovesno knjazju vygovarivaj* (Rybnikov 7:44), "He says to the prince", and *Sam on poslovečno vygovarivat*, (Rybnikov 7:85 and 119), "He says".

Line 257 is a vocative whole-line formula with a basic pattern of proper noun plus common noun describing the general category to which the person named belongs plus adjective complex. This line is similar to line 250, which is also a whole-line expression naming a person, although its syntactic pattern is different. The first part of the line, which consists of a particle or other filler plus *Vladymir knjaz'*, can also be used as an initial half-line subject, the verb of which appears in the second half of the line. It is more useful to the singer than the second part

[45] *Ibid.*, p. 69. Austerlitz calls this type of structure an "impure terrace".
[46] It should be noted that this particular hero is rarely encountered in the songs of Rjabinin that were collected by Gil'ferding.

of the line, *da stolnë-kievskoj*, which occurs only as the latter half of a line and which is always bound to an initial part expressing the idea "prince Vladimir". In other words, the first part of the line can be combined with other half-line formulas to make other whole lines, but the formula in the second part of the line cannot, or more exactly, is not combined. This adjective complex is used exclusively by Rjabinin to describe "prince Vladimir".

Line 258 was found to be a whole-line formulaic expression. The initial part of the line was also found to be formulaic, whereas the final part was a formula. The line is complicated by fillers, but the basic pattern is verb-object-adjective. If the supporting evidence in the appendix is consulted, it can be seen that the first part of the line forms a useful system consisting of a verb, plus one or more fillers which can conveniently be replaced by an indirect object (with or without particle), plus an object. We can express this system as follows:

<table>
<tr><td>No. of syllables:</td><td>4</td><td>2</td><td>3</td></tr>
<tr><td></td><td>a j nakin'-ko</td><td>tu ved'</td><td rowspan="4">} služobku</td></tr>
<tr><td></td><td>nalogaju</td><td>tobi</td></tr>
<tr><td></td><td>naložil-to</td><td>mni-ka</td></tr>
<tr><td></td><td>sosluži-tko</td><td>ètu</td></tr>
</table>

The persistence of such syntactic patterns, with particles and other line fillers to give the line formulaic stability, add additional evidence for a line that once was more dependent on syllable count than it now appears to be.[47] If the line of the text is compared with lines expressing similar ideas from songs collected by Gil'ferding from three other singers of the Kiži district, Nikolaj Fillippov Dutixov, Avram Evtixiev Čukov and Ivan Anikievič Kas'janov, then it becomes easy to see a commonness of certain formulas within the tradition of a given district:

Naložil na menja službu nesnosnuju (Dutixov 145:12)	He has charged me with an intolerable task
Ty nakin'-ko ètu službu da velikuju (Čukov 148:96)	Charge with this great task
Kak nakinul ètu službu velikuju (Čukov 148:107)	He charged with this great task
A nakinul službu da velikuju (Čukov 148:128)	He charged with a great task
Ty nakin'-ka ètu službu na Dobrynjušku (Kas'janov 157:82)	Charge Dobrynja with this task

[47] See Jakobson, "Studies in Comparative Slavic Metrics", pp. 21-66; and also "The Kernel of Comparative Slavic Literature", pp. 24-31.

On nakinul mne da službu ved' velikuju	He charged me with a great task
(Kas'janov 157:106)	

Line 259 is a whole-line formulaic expression, the initial part of which is formulaic and the final part of which is a formula. The basic pattern is adjective-noun-adjective. This line forms a formulaic couplet with the preceding line. The last two words in the preceding line, *službobku velikuju,* suggest in chiastic order *veliku službobku,* the initial formula of line 259. In addition to this, the last two words of line 259, *službobku nemaluju,* parallel the last two words of the above line, *službobku velikuju.* In our concept of formula, however, adjective-noun and noun-adjective relate to one another in terms of the lexicon of formulas and not formulas as such. Formulaic evidence could be considered as evidence for formula for the whole-line phrase, and consequently for the first part of the line, if correspondence of particles were not considered necessary. Compare the initial part of this line, *da veliku slúžobku,* with *da j veliku slúžobku* (79:275) and *i veliku slúžobku* (80:168).

Line 260 is a whole-line formulaic system, the first part of which is also formulaic and the final part of which is a formula. The pattern is a simple and useful one. It is a whole-line thirteen-syllable prepositional phrase which consists basically of a preposition plus pronominal adjective plus descriptive adjective plus object. A few varied examples of this thirteen-syllable pattern follow:

Na togo da na molódago Dobrynjužku	On young Dobrynjuška
J u togo li u molódago Dobrynjuški	In young Dobrynjuška's possession
Ko tomu on k molodomu skomorošine	To the young buffoon

The appendix can be consulted for an exhaustive listing. The frequency of use of this and other syntactic patterns will be taken up later.

According to the evidence at hand, line 261 is a formulaic expression, as is the first part of the line. The line ends in a formula. The pattern is conjunction-verb-pronoun-preposition-adverb-adjective-noun. The division of this line might seem strange, since the preposition has been separated from its object. This would have seemed a logical division had the preposition been repeated before *čisto pole*: but it is not a grammarian's logic that is at work here, rather that of a singer of epic song. I submit that there is probably a strong affinity in the singer's mind between the preposition and the adverb, since there exist such adverbial forms as *vdaleke.* It would seem then that the preposition is bound to both parts of the line, more specifically to the adverb *daleče* as well as the object *čisto pole*; that there is a tension between both the adverb and the object for the preposition. Other solutions to this problem can be found in the appendix.

Line 262 is also a whole-line formulaic expression, but the separate parts are formula. The basic pattern is preposition plus pronominal adjective plus adjective plus noun plus adjective. The whole line could actually be considered a formula if correspondence of particles and preposition line fillers were not required. There is no essential difference between the following thirteen-syllable lines (see appendix, note 40):

	slavnym ko		ko	
ko tym		goram		soročinskiim
	slavnyem		da k	

When the preposition is not repeated before *goram*, the extra syllable is provided by extending the ending of the preceding adjective, i.e., *slavnyem*, thereby making it into three syllables. Both *ko* and *da k*, which precede the final adjective, are monosyllabic, and serve only to add an extra syllable. The general pattern of this line, which is a thirteen-syllable whole-line prepositional phrase, recurs frequently. Line 260 in the chart is also a thirteen-syllable whole-line prepositional phrase, although the syntactic pattern within the line is different.

Line 263 fits into the general pattern of conjunction-verb-preposition-object-adjective. It is a whole-line formulaic subordinate clause, the first part of which is also formulaic and the latter part formula. It is worth noting that the line begins similar to line 261. The first five syllables of line 261 are occupied by conjunction-verb-pronoun, *čtoby s"ezdil on*, as are the first five syllables of line 263, *da sxodil by on*. In place of *čtoby* has been substituted *da* (verb) *by*. The verb separates these two monosyllabic elements. A different choice of conjunctions in both instances is dictated by the accent of the two verbs and their position in the initial part of the line. In both lines the first primary accent falls on the third syllable, which is the general pattern for this entire passage. Compare the beginnings of the following lines:

Kto by mog sxodit'
Da sxodi-tko ty
On velel sxodit' } vo nory.....
Ne vxodi-ko ty
Tak nel'zja vojti

The formula at the end of this line, preposition-adjective, modifies the object that immediately precedes it and that forms the end of the first part of the line. This formula parallels the final formula of the preceding line.

It should be noted that even though in each case the initial and final parts of the various lines have been taken into consideration, it is probably the whole-line expression that is the compositional device most useful to the singer in many cases.

The notes to Chart I give ample examples of an idea that can be expressed within a variety of lengths, especially within either a whole or half line. As additional evidence of this compare *vo nory vo zmeinyi*, the end of the first part plus the final part of line 263, with the following eleven-syllable whole-line expressions: *da j / da ko tym noram da ko zmeinyem/-yim* (79:19, 54 and 136), "to the dragons' lair"; *da j ot tyx ot nor on ot zmeinyix* (79:70) and *a j so tyx so nor da j so zmeinyix* (79:456), "from the dragons' lair"; and also the following thirteen-syllable expression: *ko tym slavnyim noram da ko zmeinyim* (79:225), "to the famous dragons' lair". These few examples, and many more could be added, show that a singer can express certain essential ideas just as easily within the space of the whole line as well as part of the line. He has within his technique of composition the devices necessary for making this type of adjustment, i.e., the addition or omission of initial adjectives, particles, pronouns, repeated prepositions, or expansion and contraction or morphological endings. An essential idea can be freely expanded or contracted: a half-line expression can be expanded or contracted within the limits of the half line, a whole-line expression can be expanded or contracted within the limits of the whole line; a half-line expression can be expanded into a whole line, a whole-line expression can be contracted into a half line. Syntactically, these expressions remain basically the same, differing only in syllable count.

In establishing evidence for whole- and half-line formulas and formulaic expressions in Russian oral epic poetry, the number of phrases directly supporting the results of this study would significantly increase if, when applying Parry's definition of formula to the Russian tradition, syllable count were not strictly adhered to, as was done in all cases in establishing evidence for formula and, in most cases, for formulaic expressions. The restrictions imposed by syllable count and word-for-word likeness of the "little" words (particles, repeated prepositions, etc.) limit the concept of formula in the Russian tradition. Actual concern should be with accentual groupings and basic syntactic units repeated in the same order within the whole and half line, discounting in the syntactic description the "little" words that are established as fillers, the omission of which changes not in the least the essential idea or the basic syntactic pattern. In other words, in order to define a formula in the Russian oral epic tradition more precisely, we need only adjust the concept of "the same metrical conditions" and focus on accentual configurations rather than on syllable count, since syllabic stability in the Russian tradition is only a tendency and not a law, as it is in the Yugoslav. In this study, this limited concept of the formula has been preserved, however, in the hope that the results obtained are significant, giving further indication of an epic line that at one point far back in history was stable.[48] A revised definition would be as follows:

a) A formula in the Russian oral epic tradition requires word-for-word likeness of the basic syntactic elements, excluding fillers, of at least two phrases. These basic elements must also be in the same position in the line. The rhythmical pattern of a formula consists of a flexible number of syllables organized around the same number of primary accents, three accents in the case of a whole-line formula, two if a formula is in the first half of the line, one if a formula is in the second half of the line.

b) A formulaic expression in the Russian oral epic tradition requires likeness of the basic syntactic elements, excluding fillers, of at least two phrases. These basic elements must also be in the same position in the line. Formulaic expressions must share at least one identical basic word in the same position in the line or half line. The rhythmical pattern of a formulaic expression is the same as that of a formula. The above restrictions would allow the supporting evidence to increase significantly.

IV

SOME PECULIARITIES OF COMPOSITION

The formulas of a given singer exhibit certain distinctive features under specific conditions. These features include the fixed position in a line of certain words and expressions, and the use of particular forms under different grammatical conditions, in addition to the use of formulaic substitution systems.

In the discussion of Chart I it has already been noted that certain verbs have a fixed position in the line. The verb *govorit'* appears in initial position in the line. Likewise, the verbs *vosprogovorit'*, *vozgovorit'* and *isprogovorit'* always appear in initial position. The *govorit'*-type verbs, because of accent, are not suited for position at the end of the line. On the other hand, the prefixed forms of *govarivat'*, i.e., *vygovarivat'* and *pogovarivat'* within the corpus of material examined always appear at the end of the line. The accent of the *govarivat'*-type verbs has probably played the determining role in fixing the position of these verbs. This is especially clear in verbs of this latter type, since the accent of the infinitive falls on the second from the last syllable, making it especially suited for final position in a normal line with a dactylic ending. Fixed position also seems to be felt by Rjabinin in relation to *takovy slova*, which is in position of a

[48] *Idem.*

final formula. The one exception has been discussed above in the comments to Chart I. The usefulness of this word group at the end of the line is determined, also, by its accentual pattern. The primary accent falls on the second to the last syllable, i.e., on the last syllable of *takovy*. The adverb *poslovečno*, which appears in line 256, seems to have a fixed position at the beginning of the line. Not all words or expressions, however, have a fixed position in the line. By consulting lines 253 and 254 in the chart, it is immediately clear that *za dvenad-cat' god* is just as useful to Rjabinin as an initial or as a final formula. Many other lines in the chart will give further illustration of this. See especially lines 250 and 257 and the accompanying notes in the appendix.

It is clear that a certain category of formulas, in Rjabinin's case for instance those for naming the actors and place of origin, have their own particular forms under different grammatical conditions. "Prince Vladimir" will serve as illustration. In most cases the word order is *Vladymir knjaz'* when this word group is used as the subject of a line containing also a predicate and complement, or when it is part of a vocative whole-line expression: *Vladymir knjaz' po gorenki poxažival* (81:8), "Prince Vladimir paced the chamber"; *Govoril-to emu Vladymir knjaz'* (80:831), "Prince Vladimir said to him"; *Govoril emu Vladymir knjaz' da stol'ne-kievskoj* (80:913); "Prince Vladimir of royal Kiev said to him"; or *Ty Vladymir knjaz' da stolnë-kievskoj* (79:257), "Prince Vladimir of royal Kiev". In cases other than the nominative the word order is reversed, i.e., *knjaz' Vladymir*: *Knjazju-to Vladymiru da golova srubit'* (75:30), "To cut off Vladimir's head"; *Unesla ona u knjazja u Vladymira* (79:194), "She took away prince Vladimir's"; *Stanovis'-ko suprotiv knjazja Vladymira* (75:48), "Stand up in front of prince Vladimir". The adjective complex indicating the city of origin of prince Vladimir, i.e., *stol'ne-kievskoj*, "of royal Kiev", occurs only in those lines in which "prince Vladimir" is in the nominative case. When "prince Vladimir" is in a case other than the nominative an adjective such as *slavnyj*, "famous", is used in place of *stol'na-kievskoj*: for example, *Ty Vladymir knjaz' da stol'në-kievskoj* (74:199), "Prince Vladimir of royal Kiev"; or *I zavel Vladymir knjaz' da stol'në-kievskoj* (76:43), "Prince Vladimir of royal Kiev gave" [a feast]; but *Da j ko slavnomu ko Vladymiru* (75:599), "To famous prince Vladimir's". Note also the following whole-line expression containing both *stol'në-kievskoj* and *slavnyj*, but omitting *knjaz'*: *Aj ty slavnyja Vladymir stol'në-kievskoj* (74:167), "Famous Vladimir of royal Kiev". The expression *Kalin car'*, "czar Kalin", exhibits a word-order pattern similar to that of *Vladymir knjaz'* under different grammatical conditions. In most cases the word order is *Kalin car'* when used in the nominative case: *Aj že ty sobaka da i Kalin car'* (75:80), "Oh, you infidel, czar Kalin"; or *Obošel sobaka Kalin car' naš Kiev grad* (75:144), "The infidel, czar Kalin has surrounded our city of Kiev". In cases other than the nominative the word order is different, i.e., *car' Kalin*: *U nego sobaki carja Kalina* (75:380), "The infidel, czar Kalin has". Note also the following thirteen-syllable expressions in which the patronymics *Botijanovič* and *Plenkovič* are used in the nominative, whereas the family names are used when other cases are called for: *A j korol'-ot Botijan da Botijanovič* (80:398), "King Botijan Botijanovič"; but *K korolju-to k Botijanu Botijanovu* (80:16), "To king Botijan Botijanov"; or *Aj žo molodoj Ščuriluška ty Plenkovič* (85:204), "Oh, young Ščuriluška Plenkovič"; but *Na molódogo Ščurilušku na Plënkova* (83:517), "At young Ščuriluška Plenkov". The family names in the nominative case are one syllable shorter than the corresponding patronymics and allow for the addition of morphological endings designating cases other than the nominative, at the same time preserving the same syllable count in the line. One more example will give additional illustration of this feature of formula structure. In cases involving the use of a modified form of *pir*, "feast", whether at the beginning or the end of the line, the modifier is *počesten*, "honorable", if the expression is in the accusative case. On the other hand, the modifier is some form of *čestnyj*, "honorable", if the expression is in an oblique case. Compare *Tak pošel-to Dobrynja na počesten pir* (80:804), "Dobrynja set out for the honorable feast"; or *Na počesten pir ko knjazju stal poxaživat'* (79:212), "He began to frequent the prince's honorable feast"; and *On s čestná piru idet da i ne veselo* (79:291), "He leaves the honorable feast unhappily", or *A ty eš' da p'eš' da na čestnom piru* (76:27), "You eat and drink at the honorable feast". The persistence of such patterns gives additional evidence, in my opinion, for an epic line that once was more stable than it is now.

The usefulness of formulas can be illustrated by indicating the many words that can be substituted for a key word. For example, in an initial formula like *govoril Vladymir*, any three-syllable subject can be used in place of *Vladymir*. This system can be written as follows:

	Dobrynja
	Il'ja-tot
govoril	korol'-ot
	orataj
	Sčurila

We can call *govoril Vladymir* a formulaic substitution system in which either verb or subject can be replaced by another verb or subject that fits into the same rhythmical pattern. This system belongs to a type of formulaic first half of a line in which the syntactic pattern is as important as the verbal one. Since Vladimir's role is a rather passive one, confined mostly to speaking and thinking and pacing the floor, it will be more revealing to take one of the more energetic substitutes, Dobrynja, and note a few of the three-syllable verbs that can fit into the pattern *govoril Dobrynja*. For example:

vyezžal	
vyxodil	
da j sxodil	
da xodil	Dobrynja
perešol	
priezžal	
prixodil	

The above formulas have been grouped together in "systems", a term that Parry used when studying the traditional epithets in Homer.[49] This concept of "systems", with its stress on the fact that only a group of formulas with each singer is fixed, relates to the singer's habit of reducing his manner of expression to the simplest patterns by means of analogy, by altering one formula already in use to obtain a new one.[50] A formulaic system, in other words, indicates what it is possible to change. The following examples from Rjabinin can be used to illustrate systems further. At the end of the line, instead of *vo bel šatër*, "into the white tent", one can say *v bož'ju cerkov'*, "into God's church", or *v vysok terëm*, "into the high tower". These and similar patterns can be illustrated as follows:

(2-2)
vo bel šatër (75:237)
v bož'jú cerkov' (84:168)
v vysok terëm (84:103)
v čistó pole (79:47)
v Kijan-more (80:180)
v belý ruki (75:397)
v moëj voli (79:121)
v syrú zemlju (77:326)
v zlatóm verxu (84:29)

(4-1)
vo Černigov grad (74:20)
na širokij dvor (74:207)
na posyl'noj dvor (75:41)
na počesten pir (76:70)

(3-1)
na Kiev grad (75:26)
na zólot stol (75:49)

(3-2)
vo syrú zemlju (90:34)
vo temnú ordu (80:172)
vo zlatý verxi (84:136)
vo belý ruki (75:440)
vo čistó pole (73:26)
vo syróm dubu (74:39)
vo bož'jú cerkov' (83:134)
vo Pučaj reki (79:25)
na dobróm koni (74:106)
na odno plečko (74:205)
na tvoem dvori (74:200)

[49] Milman Parry, *L'Epithète traditionnelle dans Homère* (Paris, 1928), pp. 11-15.
[50] Parry, "Studies in the Epic Technique", p. 145.

na vdovinoj dvor (80:300)	na krutój berëg (85:429)
	na odno uško (76:74)
	na syrú zemlja (77:343)
	na rezvý nožki (80:93)

The prepositional phrase, which appears in the final part of the line and which completes the sense of the whole-line expression, has been viewed as a unit in the above examples. This listing is far from exhaustive. The unit consists of a varying number of syllables which are organized around the last primary accent of the line, the dactylic ending, and is, in each case, comprised of a preposition, adjective and noun, or a preposition and nouns in apposition. A four-syllable space needs to be filled and Rjabinin, by the above examples, does so with a monosyllabic preposition, a monosyllabic adjective and a two-syllable object; or with a non-syllabic preposition, a two-syllable adjective and a two-syllable object (or two disyllabic nouns in apposition). Both rhythmical patterns can be described as 2-2. This four-syllable space can also be filled with a monosyllabic preposition, a two-syllable adjective and a monosyllabic object; or with a monosyllabic preposition and two disyllabic nouns in apposition. This rhythmical pattern is 3-1. A five-syllable space can be filled with a monosyllabic preposition, a three-syllable adjective and a monosyllabic object (or a three-syllable noun in apposition with a monosyllabic noun), with a rhythmical pattern of 4-1; or with a monosyllabic preposition, a two-syllable adjective and a two-syllable noun, with a rhythmical pattern of 3-2; or with a non-syllabic preposition, a three syllable adjective and a two-syllable noun, with a rhythmical pattern also of 3-2. This is only a sample illustration of the usefulness and relationship of certain groups of expressions. The list would expand greatly if every prepositional phrase that fits into this pattern were included. Only a few such phrases beginning with the prepositions *v* and *na* have been selected. Parry has defined the length of a system as consisting of the number of formulas that comprise it.[51] The systems discussed above will partly illustrate this concept.

The thrift of a system, thus, is judged according to the extent to which this system is free of phrases that could replace one another, phrases that have the same metrical value and express the same idea.[52] In order to illustrate a singer's thrift, some 5,400 lines of Rjabinin's songs have been examined for all instances of "they mounted their good steeds and set out". Note that when the subject is *oni*, "they", the verb "mount" is always expressed by *seli*, and the line ends in a second verb, *poexali*, "they set out". Both of these verbs are perfective. The one exception to this final verb is found in 77:300. The verb is *poroz"exalis'*,

[51] *Ibid.*, p. 86.
[52] *Idem.*

"they set out in different directions", i.e., "they parted": *J ony seli na dobryx konej da poroz' exalis'*. The following examples were found: *Oni seli na dobryx konej, poexali* (73:86, 81:242, 317; 85:405, 436; 87:252), *J oni seli na dobrýx konej, poexali* (80:345), *I oni seli na dobryx konej, poexali* (81:31). The following line could also be included with the first example: *Ony seli na dobrá konja, poexali* (84:160). The singular *dobra konja*, "good steed", is most likely a slip of the tongue on the part of the singer. The use of the singular here is probably one of association with the preceding five lines in which some form of the noun *konja*, "steed", appears in the singular, three times modified and once unmodified. When the subject is a pronoun other than *oni* or a proper name, the verb is expressed by some form of *sadit'sja*, which is an imperfective verb, and there is no second verb at the end of the line. Compare the following exhaustive listing: *On skorešen'ko sadilsja na dobra konja* (74:262; 79:188), "He quickly mounted his good steed"; *I sadilsja tut Il'ja da na dobra konja* (75:185), *I sadilsja Il'ja na dobrá konja* (75:358), *Da j sadilsja tut Il'ja da na dobra konja* (77:104), *A j sadilsja tut Il'ja on na dobra konja* (77:128) and *Da sadilsja-to Il'ja da na dobrá konja* (77:348), "Il'ja mounted his good steed"; *Kak sadilsja-to Oleša na dobra konja* (77:30), "Oleša mounted his good steed"; *Sadilsja Dunaj na dobryx konej* (81:272), "Dunaj mounted his good steed"; *Tut Dobrynjuška sadilsja na dobrá konja* (77:45), *A j sadilsja-to Dobrynja na dobra konja* (79:43), *Da j sadilsja-to Dobrynja na dobra konja* (79:343) and *A j sadilsja-to Dobrynja na dobrá konja* (80:248), "Dobrynja mounted his good steed"; *Da j sadilsja tut Skopin da na dobra konja* (88:2), "Skopin mounted his good steed"; *Kak sadilsja on da na dobra konja* (75:517), "He mounted his good steed"; *On sadilsja skoro na dobrá konja* (83:168) and *I skoren'ko-to sadilsja na dobra knoja* (80:694), "He quickly mounted his good steed"; *Ën sadilsja na dobra konja na sedlana* (84:110), "He mounted his saddled good steed"; *I sadilis' molodcy da na dobrýx konej* (75:576), *Sadilis' molodcy-to j na dobryx konej* (80:666) and *Kak sadilis' molodci-ty na dobryx konej* (85:466), "The heroes mounted their good steeds"; *A j sadites' vy da na dobryx konej* (75:264), *I sadites'-ko vy na dobrýx konej* (75:304, 330) and *Da sadites'-tko vy na dobrýx konej* (75:561), "Mount your good steeds"; *I ne budem my sadit'sja na dobrýx konej* (75f: ff284, 315, 341), "We will not mount our good steeds". Note also the following line: *On sadilsja na kobylku soloven'ku* (73:85), "He saddled his young nightingale mare".

V

STABILITY, SYNTAX AND SOUND

In the Russian *bylina* tradition, the end of the line usually is marked by a dactyl ending and a pause. The end of the line and the close of a compositional unit are synonymous. Each line represents a complete syntactic unit: a vocative expression, a subject with its modifiers, an object with its modifiers, a prepositional phrase, a subordinate clause, a complete sentence (sometimes with subject, object or auxiliary understood). The first one hundred lines of song No. 80 of Rjabinin can be divided into those in which the sense of the sentence is complete and into those in which it is incomplete.[53] The first type of complete unit is a verbless one, i.e., a whole-line vocative expression; and the second type is a complete unit with verb, i.e., a whole line containing subject (or understood subject), complement (direct object, indirect object, prepositional phrase), and complete verb (either in the strict sense or in the sense of having the verb or part of the verb understood, provided, of course, the understood part does not occur in the following line). These two types fall into the following basic patterns (the figures to the right indicate the number of times these patterns occur in the one hundred lines examined):

COMPLETE UNIT

Type I

a. without verb 6
ex: Ty, Vladymir knjaz' da stol'nё-kievskoj (line 30)

Type II

a. with verb, subject, complements 45
ex: Na čestnom piru Vladymir stal poxaživat' (line 6)

b. with verb complete but part understood 11
ex: I beret čaročku ot knjazja vo bely ruki (line 47)

total 62

The first type of incomplete unit, on the other hand, is a whole-line subject, object or prepositional phrase, and the second type is an expression containing a verb that lacks either subject or complement, that is, the subject or complement of which are expressed in another line. These types fall into the following patterns:

INCOMPLETE UNIT

Type I

a. subject 7
ex: Mólodoj-to Vasil'juško Kazimirov (line 45)

[53] Some of this terminology has been suggested to me by Barbara Widenor Maggs.

b. object 9
ex: A j molódago Dobrynjušku Mikitinca (line 60)
c. prepositional phrase 15
ex: Da vo tye dal'nie vo zemli soročinskie (line 10)

Type II

a. expression with verb incomplete, lacking subject or complement . . 7
ex: Zavodil počesten pir da j pirovan'ice (line 2)

total 38

Based on the above statistics, the complete unit with verb, subject, complements (Type II) seems to be the most useful kind of whole-line expression.

Analysis of whole-line vocative expressions and certain whole-line incomplete units, especially subjects, objects and prepositional phrases, makes it clear that in the Russian tradition it is not a matter of a longer line allowing for the frequent use of adjectives but rather the types of lines or structural units that a poet in this tradition uses to tell his tale.

Up to now, single lines and parts of lines have been examined, and little attention has been given to what precedes or follows. Inspection of a number of successive lines will indicate more clearly what types of larger units the singer in the Russian oral epic tradition deals with in line-for-line composition. A typical passage involving a series of whole-line complete units with verbs can be found among those lines examined above:

(1) Vse bogatyri za stolikom umolknuli,
(2) Vse umolknyli i priutixnuli,
(3) Kak bogatyri za stolikom-to prituljalisja,
(4) A bol'šaja-to tulítsja za serédnjuju,
(5) A serednja túlitsja za men'šuju,
(6) A ot men'šoëj ot túlicy otvetu net.
(80:19-24)

(All the heroes at table fell silent, all fell silent and calmed down. The heroes at the table huddled together; the great hid behind the lesser, the lesser hid behind the least, and from the least in rank there is no response.) In the above passage, one action is added to another. This adding style is helped or built up by syntactic parallelisms and acoustic patterns.[54] Lines one and three have the same

[54] See Roman Jakobson, "Linguistics and Poetics", *Style in Language*, ed. Thomas A. Sebeok (New York-London, 1960), pp. 369-370, in reference to those oral traditions which use grammatical parallelism to connect consecutive lines, and also his article "Poèzija grammatiki i grammatika poèzii", in *Poetics*, ed. R. Jakobson *et. al.* (Warsaw-Hague, 1961), pp. 400-402. See also Wolfgang Steinitz, *Der Parallelismus in der finnisch-karelischen Volksdichtung* (= *Folklore Fellows Communication*, No. 115, Helsinki, 1934), and Austerlitz, *op. cit.*, *passim.*

basic syntax: i.e., subject-prepositional phrase-verb. Lines four and five are also syntactically the same: i.e., subject-verb-prepositional phrase. All four lines are complete whole-line expressions with verbs. The subject consistently is in first position in the line, but the order of the verb and its complement is in one instance complement-verb and in the other verb-complement. A chiastic relationship exists. The verb at the end of the first line, *umolknuli*, is repeated at the beginning of the second line and is also balanced by a verb at the end of the second line, *priutixnuli*. There is the internal rhyme of *umolknuli* and *priutixnuli* in the second line. The verbs at the end of the second and third lines carry the same prefix *pri-*. The root of the verb at the end of the third line, *prituljalisja*, is echoed in the middle of the fourth and fifth lines by *tulitsja*. The adjective at the end of the fourth line, *serednjuju*, is repeated at the beginning of the fifth and is also balanced by an adjective at the end of the fifth line, *men'šuju*; this adjective is in turn repeated at the beginning of the sixth line. The preposition *za* in the first line is repeated in lines three, four and five.

The over-all acoustic pattern in this passage is worth special note. In the first line there is a predominance of *o*'s, *u*'s and hard and soft *l*'s, especially the combinations *-ol'-*, *-ol-* and *-ul'-*. These sounds carry over into the second line, especially the *u* sounds and the combinations *-ol-*, *-ol'-* and *-ul'-*. The third line, syntactically parallel to and almost a repetition of the first line, again repeats the *o* sound and *-ol'-*, *-ul'-* and *-al'-*. At the beginning of the fourth line there is *-ol'-* and *-ul'-* and in the middle of the fifth and sixth lines, *-ul'-*, with the *o* sound again predominating in the sixth. Hard *t*'s run through all the lines, playing an especially significant role in the sound pattern of the last line. The *o* sound that opens the first and second lines is picked up again at the end of the forth line and at the beginning and end of the fifth, and receives a kind of summing up at the end of the sixth line.

Another passage, chosen from Rjabinin's song about Il'ja Muromec and Solovej the robber, will further illustrate the importance of syntactic and acoustic patterns in building up a series of lines:

(1) A ja *exal*-to dorožkoj prjamoezžeju,
(2) Prjamoezžeju doroženkoj ja *exal mimo*-to Černigov grad,
(3) *Exal mimo* ètu Grjaz' da mimo Černuju,
(4) *Mimo* slavnu rečenku Smorodinu,
(5) *Mimo* slavnuju berezu-tu pokljapuju,
(6) *Mimo* slavnyj *exal* Levonidov krest.
(74:171-176)

(I rode along the straight road, I rode along the straight road by Černigov city,

I rode by the Black Morass, by the famous Smorodina River, by the famous leaning birch, by the glorious cross of Levonidov.)

Of special note in these six lines is the repetition of a single verb, *exal*, which appears in the first, second, third and sixth lines, and repetition of the preposition *mimo*. The verb is echoed at the end of the first line in the adjective *prjamoezžeju*; the phrase at the end of this line, *dorožkoj prjamoezžeju*, is repeated at the beginning of the second line in reverse word order and expanded by one syllable; the verb in this line is now used in conjunction with the preposition *mimo*, which reminds of the prefix *prjamo-*; the third line begins with the verb *exal*, followed by the preposition *mimo*, which is repeated later in the line; lines four, five and six begin with a repetition of this preposition, which is immediately followed by a form of the adjective *slavnyj*; *exal* is again picked up in the sixth line.

Passages built on syntactic parallelisms and acoustic patterns tend to have a comparative stability. This is especially true of couplets, which tend to exhibit clearly marked patterns. Certain couplets are used again and again with little if any change. For example:

> Otvezi-tko dani za dvenadcat' god,
> Za dvenadcat' god da s polovinoju,
> (79:253-254)

or

> I v odnoj tonkój rubašečke bez pojasa
> I v odnyx tonkix čuločikax bez čobotov,
> (80:268-269)

(In a fine smock without a girdle and in a pair of fine stockings without shoes.) For repetition of the first couplet see notes to lines 253 and 254 in the appendix (notes 13-18). For occurrences of the second couplet see 81:223-224; 84:138-139; 87:293-294. Sometimes syntactic and acoustic patterns tend to give comparative stability to a series of couplets or a number of lines that follow one another in rapid succession. Compare the following three passages describing the prowess of the hero's or heroine's horse, chosen from three different songs by Rjabinin:

> 1. Ego dobroj kon' da bogatyrskii
> Po čistu polju on stal poskakivat',
> Po celój versty on stal pomaxivat',
> Po kolenu stal v zemeljušku pogrjazyvat',
> Iz zemeljuški stal nožëk ën vyxvatyvat',
> Po sennoj kupny zemel'ki ën vyvertyval,
> Za tri vystrelu on kameški otkidyval.
> (79:377-383)

2. Ego dobroj kon' da bogatyrskii
To pošol skakat' da po čistu polju,
Po celój versty on da poskakival,
Po kolenušku v zemel'ku ën ugrjazyval,
Iz zemel'ki svoi nožen'ki vyxvatyval,
Po sennoj kupny zemeljuški vyvertyval,
Za tri vystrelu on kameški otkidyval.
(80:324-330)

3. Ona exala v pogonu po čistu polju,
A skakala na koni bogatyrskoëm
Da po slavnu rozdol'ju čistu polju;
Po celoj versty kon' poskakival,
Po kolen on v zemeljušku ugrjazyval,
On s zemeljuški nožki vyxvatyval,
Po sennoj kupnu on zemeliki vyvertyval,
Za tri výstrely kameški otkidyval.
(81:259-266)

There are certain larger groups of lines that the singer is accustomed to use often. Repetition of lines within these larger units is sometimes word-for-word the same. Often the order of the lines, or rather the sequence of ideas, is different. These clusters of formulas or lines, which are frequently found in association with one another and are recurrent, are also characteristic of oral style.[55] The examples in Chart II from "Dobrynja and Vasilij Kazimirov" will illustrate how stable such a cluster may be over a ten-year period. The first passage is from a version sung for Gil'ferding in 1871, and the second is a parallel passage in the version recited for Rybnikov about ten years earlier.

CHART II

Gil'ferding 80:249-265

Govorila tut Dobrynina-ta matuška:
Aj že svet moja ljubimaja semejuška,
Moloda Nastas juška Mikulična!
Ty čego sidiš' vo tereme v zlatom verxu?
Nad soboju li nezgoduški ne vedaëš ?
Zakataetsja-to naše krasno solnyško

[55] The problem of themes will be taken up in a separate study.

Da za ètyi za gory za vysokii,
Da j za ètyi za lesuški za temnyi,
A j s"ezžaet-to Dobrynja s široká dvora.
A j podi-tko ty skoren'ko na širok na dvor,
D j zajdi-tko ty Dobrynjuške s bela lica,
Pódojdi k nego ko pravomu ko stremjačke,
Govori-tko ty Dobryne ne s upadkoju,
Poprosi-tko u molódago Dobrynjuški:
On daleče l' edet, kudy put' deržit?
Skoro ždat' velit nam, kogdy dóžidat'?
Nam kogda j velit v okošečko posmatrivat'?

Rybnikov 8:205-221

Govorit ego roditel'-matuška,
Čestna vdova Ofim'ja Aleksandrovna:
Aj že ty, moja ljubeznaja semejuška,
Moloda Nastas ja doč' Mikulična!
Ty čego sidiš' vo tereme v zlatom verxu?
Ali nad soboj nevzgoduški ne vedaeš'?
Zakataetsja-to naše krasnoe solnyško
Za èti za lesuški za temnye
I za tyja za gory za vysokija:
S"ezžaet-to Dobrynja s široka dvora.
Podi-ka ty skoren'ko na širokij dovr,
Zajdi-ka ty k Dobryne s bela ličika,
Podojdi k nemu ko pravomu ko stremjačku bulatnemu,
Govori-ka ty Dobryne ne s upadkoju:
Kuda, Dobrynja, edeš', kuda put' deržiš',
Skoro l' ždat' nam veliš', kogda sožidat',
Kogda veliš' v okošečko posmatrivat'?

In Chart II solid underlinings indicate exact or nearly exact repetition in the two passages. Spelling variants and particles have been ignored. A broken underlining indicates close correspondence of words and phrases.

A passage selected from the beginning of this same song will show considerably less repetition and a greater difference in the order of the lines. As in Chart II, the first passage is from the Gil'ferding text and the second passage is from the text in the Rybnikov collection.

CHART III

Gil'ferding 80:6-18

Na čestnom piru Vladymir stal poxaživat':
— Aj, že vy moi da knjazja bojara,
Sil'ni russkie mogučie bogatyrja!
Zadolžal-to ja ko kórolju-to Botijanu Botijanovu
Da vo tye dal'nie vo zemli soročinskie
Da za starye-to gody i za nonešnij,
Ispolna ja gosudarju za dvenadcat' god,
Za dvenadcat' god da j s polovinoju.
Kogo že mne poslat' tudy otvézti dani vyxody
Za dvenadcat' god da j s polovinoju
K korolju-to k Botijanu Botijanovu,
A dvenadcat' lebedej, dvenadcat' krečenej,
A 'šče gramotu da i povinnuju?

Rybnikov 8:13-27

Sam Vladimir knjaz' po gorenke poxaživat,
Poslovečno gosudarin vygovarivat:
— Krasnoe solnyško na večere,
Xoroš česten pir idet na vesele,
I vse dobry molodcy porasxvastalis';
A mne, knjazju Vladimiru, čem budet poxvastati?
Kogo poslat', bratcy, iz vas povyexat'
Vo dal'nija vo zemli v Soročinskija
K korolju-to Butejanu Butejanovu:
Otnesti-to nadot' dani-vyxody
Za starye gody i za nynešni,
I za vse vremena za dosjulešny,
Ispolna gosudarju za dvenadcat' let,
Dvenadcat' lebedej i dvenadcat' krečenej,
I otvezti ošče gramotu povinuju?

These lines correspond as follows: G 10-R 20; G 11-R 23; G 12-R 25; G 14-cf. R 19 and 22; G 16-R 21; G 17-R 26; G 18-R 27.

Another passage selected from three singings of one song, "Dunaj", by Rjabinin will show considerable stability. The first passage is from the version of "Dunaj" sing for Gil'ferding in Kiži in June of 1871, the second is a fragment of

819-820; 81:8-9, 65-66, 350-351; 85:191-192. See the following line for all occurrences of the verbs "vosprogovorit'", "vozgovorit'" and "isprogovorit'": 81:226; 84:61, 97, 145; 85:383: 87:24, 80.

2. See note 1. Cf. also 74:238 (Il'ja-tot) and 88:28 (Skopin-to). For pattern see note 1.

3. See note 1 and 73:49, 59, 76, 160, 162, 172; 74:22, 71, 99, 104, 110, 162, 177, 198, 209; 75:108, 115, 141, 311, 399, 446, 448, 457, 463, 471, 610: 76:54, 66, 88, 115; 77:37, 74, 97, 121, 268; 78:56, 58, 67, 77, 84; 79:119, 242, 272, 348; 80:53, 161, 234, 279, 387, 472, 495, 560, 644, 780, 901, 920, 930, 944, 956, 968, 976, 991, 1004; 81:180, 198, 206, 226, 305, 367, 399, 443; 83:67, 149, 161; 84:24, 61, 83, 97, 145; 85:98, 170, 203, 228, 377, 430, 498, 506, 540; 87:74, 225, 284; 88:36; 90:131, 210. Cf. 85:192 (takovy slova Ščurilo pogovarivat). For pattern see note 1.

4. 75:464; 76:67. Pattern: particle-pronoun filler (2), adjective (3), noun (2), particle-noun in apposition (3), noun in apposition (3). Cf. also the following: 74:163, 75:142; 472; 77:92 (aj že staryja kazak da Il'ja Muromec); 74:218, 226; 76:116 (aj že staryja kazak ty Il'ja Muromec); 75:395 (aj šče staryja kazak da Il'ja Muromec); 74:68 (a j kak staryj-ot kazak da Il'ja Muromec); 74:78, 143 and 75:167, 351 (a j tut staryja kazak da Il'ja Muromec); 74:261 (a j tut staroj-ot kazak da Il'ja Muromec); 77:284 (a j tut staryj-ot kazak da Il'ja Muromec); 76:28 (a kak staroj-ot kazak da Il'ja Muromec); 75:524 (a tut staryj-ot kazak da Il'ja Muromec); 77:200 (a šče staryja kazak-ot Il'ja Muromec); 75:125 (da tam staryja kazak da Il'ja Muromec); 75:516 (ešče staryja kazak da Il'ja Muromec); 76:94 and 77:302 (ešče staryj-ot kazak da Il'ja Muromec); 75:151 (tak tut staryja kazak da Il'ja Muromec). Note also the following 11- and 12-syllable whole-line formula and formulaic expressions: 75:235, 248, 282, 313, 339, 501 and 76:90 (staryja kazak da Il'ja Muromec); 75:493 (tut staroj kazak da Il'ja Muromec); 75:199 (tut staryja kazak da Il'ja Muromec). Cf. also the following examples: 74:157 (govoril-to staryja kazak da Il'ja Muromec); 74:160 (est' ja staryja kazak da Il'ja Muromec); 75:117 (ved' est' živ-to staryja kazak da Il'ja Muromec); 75:260 (govoril im staryja kazak da Il'ja Muromec); 75:294, 325 (govorit-to staryja kazak da Il'ja Muromec); 75:581 (usmotrel li staryja kazak da Il'ja Muromec); 75:596 (govoril staroj kazak da Il'ja Muromec); 77:20 (govoril tut staryja kazak da Il'ja Muromec); 77:84 (govorit im staryj kazak da Il'ja Muromec). All of the above examples have concerned the nominative case. Compare the following whole-line expressions involving other cases: 76:6 (starago kazaka Il'ji Muromca); 75:498 (a u starago kazaka Il'ji Muromca); 85:387 (a j a starago kazaka Il'ji Muromca); 77:232 (da j u starago kazaka Il'ji Muromca); 77:255 (tut u starago kazaka Il'ji Muromca); 75:10 (tut u starago kazaka Il'ju Muromca); 76:48 (togo starogo kazaka Il'ju Muromca); 76:7

(staromu kazaku Il'je Muromcu); 76:65 (staromu kazake Il'ji Muromcu); 76:56 (k staromu kazake Il'ji Muromcu); 75:24 (tomu staromu kazaku Il'je Muromcu). Note also the following: 75:109 (... starogo kazaka Il'ji Muromca); 74:215 (... u starogo kazaka Il'ji Muromca); 75:497 (... starogo kazaka Il'ju Muromca); 80:190 (... v starogo kazaka Il'ju Muromca); 75:459 (... da starogo kazaka Il'ju Muromca); 75:2, 443 (... na starogo kazaka Il'ju Muromca); 75:515 (... k staromu kazaku Il'je Muromcu); 76:79, 113 (... k staromu kazake k Il'je Muromcu); 77:185 (... ko staromu kazake k Il'je Muromcu); 79:248 (... k staromu k kazake k Il'ji Muromcu).

5. See note 4. For pattern see note 4.

6. See note 4. For pattern see note 4.

7. Pattern: particle-pronoun filler (2), imperative-particle (3), preposition-object-particle (4), preposition-adjective (4). Cf. 79:244 (a j ty s"ezdi-tko vo zemlju v politovskuju).

8. See note 7. Cf. 79:276 (a j ty s"exdi-tko vo daleče). Cf. also 79:52 (emu s"ezditi vo daleče), 226 (kto by mog sxodit' vo nory), 230 (ne mogut-to s"ezditi vo daleče), 261 (čtoby s"ezdil on v daleče), 263 (da sxodil by on vo nory), 273 (a j ty s"ezdi-tko vo daleče), 278 (da sxodi-tko ty vo nory), 310 (velel s"ezdit' mni vo daleče), 312 (on velel sxodit' vo nory), 434 (ne s"ezžat'sja b nam v rozdol'ice). We might also include the following variations on a theme, i.e., "motion" verbs: 79:287 (vyxodil-to on [z]za stolikov); 74:14 (a pod"exal kak ko siluške); 79:76 (on povyexal v rozdol'ico); 80:145 (to my s"edemsja v rozdol'ici), 353 (da j priexali vo zemli), 372 (i pošli ony v polaty), 650 (i pojdemte-tko v polaty), 654 (da j prošli oni v polaty), 674 (priezžali-to ko dubu-to), 723 (da sxodit' velel ko knjazju), 741 and 768 (da scodit' velel v polaty), 743 and 770 (sxodit' velel na pogreb); 84:124 (to on poexal po rozdol'ju), 126 (on pod"exal kak ko teremu); 85:6 (da povyexat' na matušku), 58 (on poexal po rozdol'icu), 66 (da povyexal na matušku), 248 (podošla ona k konju), 267 (podxodil-to on k konju), 409 (poraz"exali s rozdol'ica), 472 (da i zaexali na goru). For pattern see note 7.

9. See note 7. Cf. 80:10 and 353 (vo zemli soročinskie/ii), 171 (vo zemli v soročinskii). For pattern see note 7.

10. Pattern: particle-preposition-pronoun (4), object (3), pronoun filler (1), imperative (2), preposition-adjective (3), Cf. line 251 in the text and also 86:6 (ko tomu-to zašli k korolju k litovskomu). Compare lines 251 and 252 in the text with 79:244 and 245 (a j ty s"ezdi-tko vo zemlju v politovskomu,/k korolju-to k Čubadeju k politovskomu). Note that 79:244-245, 251-252 and 86:8 are 13-syllable lines. Note also 89:26 (i zašol-to k korolju k litovskomu).

11. See note 10. Cf. also 81:71, 79 and 176 (u togo u korolja) and 80:381 and 81:163 (samomy-to korolju). For pattern see note 10.

12. See note 10. For pattern see note 10.

13. 79:246. Pattern: imperative-particle (4), direct object (2), preposition-object (4), dependent genitive (1). Cf. 80:36 (j otvezet en dani), 56 (a j vezu ja dani), 79 (i vezu ja dani). Cf. the following longer expressions: 80:103 (da vezu ja èti dani), 114 (prinosi-tko ty nam dani), 169 (otvesti-to nado dani), 390 (privezli kak k tobi dani), 652 (polučat'-to budem dani), 655 (prinosil korol' im dani).

14. See note 13. Cf. 79:239 (u menja-to dani) and 80:366 (da beret-to dani). For pattern see note 13.

15. See note 13 and also 80:12. See note 16 for occurrence of this formula at the beginning of the line. For pattern see note 13.

16. 79:240, 254; 80:104, 115, 170, 319. Pattern: preposition-object (4), dependent genitive (1), particle-preposition-object (6). Cf. the following 11-syllable lines in which only the monosyllabic particles differ: 79:247 and 80:13, 15, 37, 80 (da j) and 80:57 (i).

17. See note 16 and also note 15. For pattern see note 16.

18. See note 16. For pattern see note 16.

19. Pattern: adverb filler (1), subject (4), subject (4), preposition-object (4), verb (4). Cf. 85:191 (a j Ščuriluško-tot Plenkovič po gorenke en Permin da poxaživat'), 111 (stal po gorenki Vasil'juško poxaživat'); 79:234 (a j po gorenki da knjaz' poxažival); 85:453 (da po gorenke on da poxaživat); 79:217 (on po gorenke da i poxaživat). See note 1.

20. See note 19. Cf. this expression at the end of the line: 77:27 (govoril-to tut Olešenka Grigor'evič). For pattern see note 19.

21. See note 19 and also 90:819. Cf. 81:8 (poxažival) and 90:49 (po gorodu poxaživat'). Cf. also 81:350 (poxaživati). For pattern see note 19.

22. Pattern: adverb (4), indirect object (2), verb (5). Cf. 80:112 (poslovečno stal on knjazju vygovarivat') and 81:66 (poslovečno knjazju en vygovarivaet). Cf. also 79:218 and 235 (poslovečno gosudar' on vygovarival). See note 1.

23. See note 22. For pattern see note 22.

24. See note 22 and 85:454. Cf. 81:9 (vygovarival). For pattern see note 22.

25. 74:199, 239; 75:51; 80:30, 76, 866, 902, 921, 945; 83:68; 85:99, 141, 324. Pattern: pronoun filler (1), noun (3), noun in apposition (1), particle-adjective complex (6). Cf. the following whole lines which are the same as the above except for the initial monosyllabic word: 80:1 (a); 76:26 (aj); 85:455 (a j); 75:1 (kak); 76:103 and 83:171 (tam); 74:244 and 75:65, 605, and also 80:87, 166 and 85:328 (to); 75:75, 104 and 76:40, 45, 111, and also 79:270 and 80:61 (tut). Cf. also the following whole lines which are syllabically the same as the above, but which differ from the formula in the text in the first and sixth syllables: 74:259, 80:940, 83:29 (a...-ot...); 80:957 (a...ty...); 79:307 (a j ...-ot...); 80:821

(aj...ty...); 77:75; 85:315 (kak...-ot...); 80:856 (sam...-ot...); 74:203; 79:216 (to...-ot...); 75:119, 233; 80:120 (tut...-ot...); 80:99, 113, 814 and 85:441 (ty...naš...). Note also the following whole-line expressions with a different syllabic count: 81:130 (Vladymir knjaz' stol'ne-kievskoj), 32 (Vladymir knjaz' i stol'ne-kievskoj), 1, 103 and 320 (a Vladymir knjaz' stol'ne-kievskoj), 67 and 117 (ty Vladymir knjaz' stol'ne-kievskoj), 96 (a Vladymir ty knjaz' stol'ne-kievskoj); 80:38 (i tut Vladymir knjaz'-ot stol'ne-kievskoj); 75:611 and 80:54 (aj že ty Vladymir knjaz' da/de stol'ne-kievskoj). There are a number of whole-line expressions in which "Vladymir knjaz' da stol'ne-kievskoj" is used in conjunction with a verb at the beginning of the line: 74:217 (govoril-to kak...); 75:9 (a čto posadil...); 76:43 (i zavel...); 80:913, 83:5, 85:346 (govoril emu...). Cf. another group of expressions that adds an adjective at the beginning of the line: 80:931, 83:14 (aj ty slavnyja...); 80:1011 (aj ty slavnyj...). See also 75:116 (aj ty batjuško Vladymir knjaz' naš stol'ne-kievskoj).

26. See note 25. Cf. also 75:25 (a Vladymir knjaz'); 85:139 (to Vladymir knjaz'), 190 (da Vladymir knjaz'); 87:40 (a j Vladymir knjaz' on). Cf. also the 4-syllable formula "Vladymir knjaz'": 81:8 and 85:164. For pattern see note 25.

27. See note 25. For pattern see note 25.

28. Pattern: particle-particle (1), imperative-particle (3), pronoun filler-particle (2), direct object (3), adjective (4). Cf. 79:274 (nalogaju tobi služobku velikuju), 308 (naložil-to mni-ka služobky velikuju), 333 (sosluži-tko èto služebku velikuju), and also 80:167 (naložil en služobku velikuju). Note also the following: 75:76-77 (vidit: est' èto delo nemaloe,/ a nemalo delo-to velikoe).

29. See note 28. For pattern see note 28.

30. See note 28. Note also 74:14 (ko siluške velikoej). For pattern see note 28.

31. Pattern: particle-adjective (4), direct object (3), adjective (4). Cf. 79:275 (da j...nemalen'ku) and 80:168 (i...). Cf. also 85:400 (a j velik zaklad da to ne malyj) and 79:309 (a j velikuju mne služobku ne maluju). See also note 28.

32. See note 31. For pattern see note 31.

33. See note 31. For pattern see note 31.

34. Pattern: preposition-pronoun-particle (4), preposition-adjective (5), object (4). Cf. 79:127, 145 and 80:603 (j u togo li u molodogo Dobrynjuški). See also 80:858 (ko tomu on k molodomu skomorošine); 78:50 (da so ètoj so tatarskoj so slobodušk i); 78:51 (da so ètoj li s Marinkinoj so ulicy); 79:10, 209, 294 (ko svoej ko rodnoju ko matuški); 79:149 (po tomu po drugomu po berežku); 79:469 (vo ego vo belyj vo ručuški); 80:130 (so svoix-to on so belyix so ručušek); 80:422 (en so toju so velikoj so gorjačnosti); 80:520 (vo tom li-to da vo tupom koncu); 85:153 (u moej-to rodnoju u matuški); 85:181 (u moej-li to u rodnoju u matuški); 85-213 (ko svoeg ko rodnoju ko matuške); 85:237 (a j na

svoj-to on da na sirokij dvor); 85:474 (a vo tom li to vo slavnom Volyn-gorode); 87:15 (da na toj li kak na tixoej zaberegi); 87:16 (da na tom zelenoem na zatres'i).

35. See note 34. For pattern see note 34.

36. See note 34. For pattern see note 34.

37. Pattern: conjunction-verb (4), subject (1), preposition-adverb (3), adjective (2), object (2). Cf. lines 251 and 263 of text the text (notes 7 above and 43 below) and 79:52 (emu s"ezditi vo daleče čisto pole), 230 (ne mogut-to s"ezditi vo daleče čisto pole), 276 (a j ty s"ezdi-tko vo daleče vo čisto pole) 310 (velel s"ezdit' mni vo daleče v čisto pole).

38. See note 37 and also notes 7 and 8 above. For pattern see note 37.

39. See note 37. Cf. 79:134 (daleče v čisto pole). Cf. also 74:87 (po slavnu po čistu polju), 134 (po slavnomu čistu polju); 75:285, 316 and 342 (vo slavno vo čisto pole). For pattern see note 37.

40. Pattern: preposition-pronoun (2), adjective (2), preposition-object (3), particle-adjective (6). Cf. the following 13-syllable lines: 79:231, 277 (ko tym slavnym ko goram ko soročinskiim); 79:311 (ko tym slavnym ko goram da k soročinskiim); 79:18 (ko tym slavnyem goram da ka soročinskiim); 79:244 (ko tym slavnyim goram da soročinskiim); 79:69 (ot tyx slavnyix ot gor ot soročinskiix). Cf. also 79:135 (da j ko tym ona goram ko soročinskiim) and 80:237 (da vo toju vo zemli vo soročinskoju). Note also 79:225 (ko tym slavnyim noram da ko zmeinyim).

41. See note 40. Cf. 79:24 (ko toj slavnoju). For pattern see note 40.

42. See note 40 and 79:437, Cf. 79:61 and 360 (da k soročinskiem/iim). For pattern see note 40.

43. Pattern: particle-verb-particle (4), pronoun subject (1), preposition-object (3), preposition-adjective (5). Cf. lines 251 and 261 of the text (notes 7 and 37). Cf. also 79:226 (kto by mog sxodit' vo nory vo zmeinyi), 278 (da sxodi-tko vo nory vo zmeinyja), 312 (on velel sxodit' vo nory vo zmeinyi). Note also 79:21 (ne vxodi-ko ty vo nory v zmeinye), 402 (tak nel'zja vojti vo nory vo zmeinyi), 452 (vyxodite-tko so nor by so zmeinyix).

44. See note 43 and also note 8. For pattern see note 43.

45. See note 43 and 79:398, 402, 406. Cf. 79:70 (ot nor on ot zmeinyix), 407 (na nory na zmeinyi), 423 (so nor da so zmeinyix), 449 (iz nor on iz zmeinyix), 452 (so nor vy so zmeinyix), 456 (so nor da j so zmeinyix). Cf. also 79:387 and 399 (ko noram da ko zmeinyim), 415 and 427 (po noram da po zmeinyim). For pattern see note 43.

The following passages were mechanically repeated by the recording scribe and do not necessarily represent, in my opinion, an actual word-for-word repetition on the part of the singer: 75:311-327 = 280-296, 328-335 = 302-309, 336-349 =

279-292, 422-428 = 411-417; 76:35-39 = 20-24; 80:764-777 = 737-750, 877-881 = 845-850; 90:108-115 = 10-17, 141-144 = 52-55, 166-171 = 86-91, 175-250 = 96-171.

GRIBOEDOV AS A DIPLOMAT

D. P. COSTELLO

University of Manchester

In diplomacy, as in literature, Griboedov was an amateur of genius. His vocation was not to be a professional diplomatist, the efficient executant of a policy laid down by some Capodistria or Nesselrode. For this he was too independent a spirit. What career might have permitted him to realize to the full his extraordinary natural capacity one can only guess. Not literature, any more than diplomacy: neither the Arakčeev period of Alexander's reign nor the ensuing despotism of Nicholas provided the conditions in which Griboedov's literary talents could unfold freely. As he put it himself, "it is a torment to be a passionate dreamer in the land of eternal snows".[1]

It was not of his free choice that he embraced a diplomatic career. For more than a year after his retirement from military service he had led a gay and aimless existence in St. Petersburg, before entering the State College of Foreign Affairs. What moved him to take this step is not known for sure. It may be relevant that it was in March 1817 that a serious dispute began between Griboedov's mother and the 780 "souls" of her Kostroma estate;[2] Nastasja Fedorovna's circumstances were not good; and she may well have thought this a good moment for her son to embark on a regular career. (Nastasja Fedorovna throughout her son's adult life was constantly urging him to make good in the service of the state, and we hear an echo of her admonitions in Famusov's: "А главное, поди-тка, послужи").

Nevertheless, to be registered on the rolls of the College of Foreign Affairs did not commit a young man to much. Puškin, who was on the roll at the same time as Griboedov, seems to have done practically no work, and, to judge from his correspondence, Griboedov's life was not changed appreciably by his entrance into the College. His effective debut in the foreign service took place a year later.

In November 1817 he was involved in a double duel which made a consider-

[1] Letter to S. N. Begičev of 9 December 1826. Cf. *Polnoe sobranie sočinenij A. S. Griboedova*, ed. N. K. Piksanov and I. A. Šljapkin (Petrograd, 1911-1917), III, 196.

[2] N. K. Piksanov, "Griboedov i krepostnye raby," *Griboedov* (Leningrad, 1934), p. 85 and ff.

able scandal at the time. The principals were, on the one side, Count Zavadovskij and Griboedov; on the other, an officer of the Horse Guards, Šeremetev, and the future Decembrist, Jakubovič; the subject — a dispute over the affections of the ballerina Istomina. After Zavadovskij and Šeremetev had fired off, Griboedov and Jakubovič, in their turn, were to shoot. Šeremetev was mortally wounded in the first stage of the duel, and the second was postponed. Of the three survivors, Zavadovskij was, apparently, compelled to leave Russia; Jakubovič was expelled from the Life Guards and sent to active service in the Caucasus. Griboedov alone appears to have gone unpunished. It has been suggested that his appointment in 1818 to the first permanent Russian legation to Persia was a kind of honourable exile, Nastasja Fedorovna's connections in court circles having on this occasion, as later, in 1826, been effective in saving her son from more severe punishment. This, however, is only surmise. Even without the affair of the duel, Nastasja Fedorovna would have been delighted to see her son appointed to a post which carried social prestige and a fair salary.

A. S. Sturdza, who at this time occupied an important post in the Ministry of Foreign Affairs, recalled many years later that Griboedov was offered the choice between diplomatic posts in Persia and in the United States, and that he hesitated some time before making up his mind.[3] The story may be true, although I have seen no other evidence for it. Sturdza, incidentally, seems to have made a slip in saying that the choice offered Griboedov was between Philadelphia and Tehran: Philadelphia had ceased in 1800 to be the capital of the United States.

In his extant correspondence Griboedov first mentions his impending diplomatic appointment in a letter to his friend Begičev, dated 15 April 1818,[4] which shows that from the beginning the diplomatic career did not appeal to him. After some talk of his comedy, "Feigned Infidelity", he writes: "But enough of 'Feigned Infidelity'; now I'll reveal my unfeigned sorrow. Imagine, they are firmly determined to send me — where to, would you think? To Persia, and for a long time. Try as I may to get out of it, nothing does any good."[5] He goes on to describe his reception by Nesselrode, who urged him to accept the Tehran post, while he, Griboedov, raised objections and posed conditions. In a letter of 30 April, i.e., just over a fortnight later, he writes to Begičev of his departure to Persia as of something fixed and settled.[6] His formal appointment did not take place until 14 July.

The head of the Russian legation to Persia was Simon Mazarovič, a Venetian

[3] *A. S. Griboedov v vospominanijax sovremennikov* (Moscow, 1929), p. 331.
[4] All dates given are in Old Style, except where otherwise indicated.
[5] *Polnoe sobranie*, III, 128.
[6] *Ibid.*, p. 130.

nobleman, and a good illustration of the non-national character of Russian diplomacy at a time when the First Secretary of State for Foreign Affairs was the Westphalian Nesselrode, and the Second the Greek Capodistria. Mazarovič himself did not take Russian citizenship until 1836. Griboedov's relations with his chief appear from his correspondence to have been tolerably good — in the sense, that is, that they do not seem to have disliked each other personally, although Griboedov's attitude to Mazarovič became in the course of time one of more or less good-natured contempt. Mazarovič proved to be a rather weak champion of Russia's interests; moreover, the fact that the head of the Russian legation was a foreigner must have been a source of irritation to one who had as lively a sense of national dignity as Griboedov. As an ex-officer of the period of Russia's triumph over Napoleon, he had his full share of the national pride which marked the men of his class and generation and which formed so prominent an element in the outlook of the Decembrists and of Čackij.

Mazarovič and his suite were obliged to wait in Tiflis until the end of November, when the Governor of the Caucasus, Ermolov, who was in immediate control of Russia's diplomatic relations with Persia, returned from campaigning against the mountain tribesmen. They did not leave for Persia until 28 January 1819.

On the way to Tehran Griboedov's national pride was flattered by one incident: in Erivan (still a Persian possession) the Russians were actually given *chairs* to sit on. He wrote to Begičev on 3 February: "In front louts carried chairs into the house allotted to us, and I was cheered up by the fact. This amounted to a particular mark of distinction for us Russians: whereas the English meekly bend their knees and sit on the floor with their shoes off, the best way they can, we sit cheerfully on our lofty seats and trample the precious Persian carpets with our thick soles."[7]

The arrival of the Russians in Tehran was duly reported to the British Secretary of State on 1 April (N.S.) by the Chargé d'Affaires, Henry Willock. In the same despatch Willock reported that Mazarovič had been unwilling to yield him the *pas* at the ceremonies of the Persian New Year: "It was agreed that we should be so placed at court that neither should have the pre-eminence, to which I offered no objection. The Shah was particular in so dividing his attentions as to give offence to neither party." Mazarovič, said Willock, is "very moderate in his language to the Ministers. He studiously avoids all subjects tending to lead to discussions. ... The Shah, it seems, feels uneasy when too closely watched by his neighbours the Russians. His Majesty has signified to M. Mazarowitch that he must reside at Tabriz, and this gentleman has consented without a struggle, to forego the great advantage in the promotion of the

[7] *Ibid.*, p. 45.

interests of his sovereign, of having easy access to the Shah and his ministers. This step will involve him in much future trouble, as the Prince Royal and Mirza Bozurg are extremely jealous and captious in all transactions connected with their relations with Russia."[8]

In a private letter of the same date to James Morier, Willock wrote: "Mazarovitch ... courts popularity very much amongst the great, rather too much so, to be successful; this will not do in Persia, a too humble deportment is apt rather to inspire contempt than respect. He is a very pleasant agreeable man and we live on the best of terms. The King does not like to be too closely observed by the Russians, and he has desired M. Mazarovitch to reside at Tabriz, and to my great astonishment he has made no objection. He will now fall into the hands of Mirza Bozurg and meet constant opposition".[9]

A despatch of Willock's dated 14 June (N.S.) 1819, affords another glimpse of Mazarovič as a feeble negotiator:

"Since I addressed Your Lordship in April M.M. has canvassed with the Persian premier two points with some warmth and equal ill success — viz., the welcome accorded to the Georgian Prince Alexander by 'Abbas Mirza, and the establishment of Russian consulates in Persia for the protection of commerce."

The Persian Prime Minister replied to Mazarovič's representations that under the terms of the existing treaty both countries were allowed to protect fugitives from each other's territory, and that Persia intended to avail itself of this right in the case of Prince Alexander. As for the consulates, "the Sudr observed, that commerce was a branch of policy exclusively interesting to merchants and only worthy the attention of those who professed its pursuit; that it never entered His Persian Majesty's consideration; the commercial relations of the two countries were conducted with propriety and justice, and that adequate protection was already afforded by H.M. to ensure the security of the trade. ... M. Mazarowitch has not again touched on these subject, and I conclude he awaits further instructions from General Yermolow."[10]

One can appreciate that Mazarovič's ineffectiveness, which Willock himself describes with something like contempt, must have aroused the indignation of his Secretary. Apparently Griboedov made no secret of his feelings. On 3 October (N.S.) 1819 Willock wrote:

"M. Gribayedoff ... contrasts the power of one State with the weakness of the other, which he says ought to be productive to Russia of a commanding influence over the Councils of Persia, whereas M. Mazarowitch, of whom he speaks very indecorously, does not sufficiently uphold the dignity of his Sovereign, and is

[8] Foreign Office Records, 60/14.
[9] F.O. 60/15.
[10] *Ibid.*

too moderate and submissive in all his communications with the Persian Ministers."[11] I would emphasize again, that according to Willock's own testimony, Griboedov was right as against his chief.

Immediately on the return of the Russian legation to Tabriz after nearly six months at the court of the Shah, Griboedov was offered, and seized, the first opportunity to display that inflexible determination with which throughout his diplomatic career he was to defend the interests and dignity of his country.

The élite of the Persian army at this time was the corps of Russian prisoners of war and deserters, which was constantly being replenished by defections from the Russian army in Georgia. Conditions of service in Russia were rigorous; in Persia the Russian deserters were given relatively good, though unpunctual, pay and easy conditions of life. Nevertheless, some of them came in time to repent their backsliding and to endeavour to return home, while the Persian authorities, naturally, did their best to keep them. The arrival of the Russian legation in Tabriz on 20 August 1819 at once aroused the hopes of these deserters, some of whom, as we learn from Willock's despatch of 3 October (N.S.),[12] took refuge in the legation premises. As early as 23 August Griboedov noted in his diary: "Negotiations on behalf of the prisoners. Rage and Sorrow." The entry for the following day runs: "Idem. Anonymous letters. If I have to, I'll lay down my life for my unhappy countrymen."[13] This is the language of intense feeling; Griboedov was to show by his actions in the following weeks that it was sincere.

On two occasions he brought a party of repentant soldiers to the Crown Prince, ʿAbbas, to arrange for their return to Russia. Griboedov's own account of the second interview with Prince ʿAbbas is extant. He states that the Prince took each soldier aside and spoke to him separately in an endeavour to persuade them to stay in Persia, but without success. Then followed a long and fruitless argument, the Persian Prince twisting and turning like a hare, the Russian diplomatist keeping hard on his traces. At one point ʿAbbas asked: "Why don't you behave like the English? They are quiet and mild. I am very well pleased with them", to which Griboedov replied: "We don't model ourselves upon the English or upon anyone else." In the end ʿAbbas agreed to allow the 70 soldiers whom Griboedov had brought with him to return to Russia. Griboedov then asked him to release a number of others who, he stated, had expressed a desire to return to Russia but who were still being held by the Persians. He had previously communicated a list of these men to the Crown Prince. ʿAbbas called in an officer of the corps of deserters, Samson Makincev, to check the list. At this,

[11] F.O. 60/14.
[12] *Ibid.*
[13] *Polnoe sobranie*, III, 61.

Griboedov says, "I could hold out no longer, but declared that not only was it shameful to have this scoundrel in one's entourage, but it was still more shameful to produce him before a Russian officer and gentleman. What would the Naib-Sultan say, I asked, If our Emperor were to send a runaway Armenian from Persia to negotiate with him? [He replied] 'He is my *noukar*'.[14] 'I don't care if he is your general; to me he is a blackguard and a knave, and I should not have to set eyes on him'. At this he got angry and started to say all sorts of silly things to me; I paid him in sort twice over; he decided he didn't want to have anything more to do with me, and we parted."[15]

This last episode is described more decorously by Willock:

"M. Gribayedoff expressed himself indignant at being requested to confer with a person who was a traitor to The Emperor of Russia, and who had deserted from H.I.M.'s service, and used such strong and, the Persians say, such improper language that the Prince ordered him from his presence, complained to M. Mazarowietz of his conduct, and requested that he might never again be charged with any communication to him or to his Ministers."[16]

Such behavior on the part of a diplomatist of 24 years of age to the Heir Apparent of the monarch to whom he was accredited may appear crude and unprofessional. It would be naive, however, to imagine that it is the duty of a diplomatist always to be diplomatic, and in fact Griboedov won his point. Early in September he left Tabriz in order personally to conduct a party of 160 soldiers back to Russian territory (at the same time to seek new instructions from Ermolov). He performed this mission successfully and won Ermolov's approval for his manner of proceeding, though he received a reprimand from the Ministry, which considered that a "a diplomatic official ought ought not to have behaved in such a fashion."[17]

If one may be permitted to express a judgment, Ermolov and Griboedov were right in this matter and the Russian Ministry wrong. The diplomatic representatives of Russia could not decently have allowed their countrymen to be detained in Persia against their will and in contravention of the Treaty of Gulistan. As to the roughness of Griboedov's proceedings, we have Willock's testimony to the fact that it did not pay to be accommodating in one's dealings with the Persians.

Griboedov did not conceal from his European acquaintances, whether Russian or English, his contempt for the Persian grandees with whom he was thrown into contact. Thus, in a letter to a friend in Tiflis, à propos of Mirza

14 Servant (Persian).
15 *Polnoe sobranie*, III, 61-64.
16 F.O. 60/14.
17 *Polnoe sobranie*, I, xl.

Mas'ud, Prince 'Abbas' Minister of State, who had gone on a mission to Ermolov, he wrote: "Tell me about Mirza Mas'ud, what sort of reception the Commander-in-chief gave him. If he orders him to receive a beating, this will greatly improve our standing in Persia."[18] Willock had written, in his despatch of 3 October (N.S.): "M. Gribayedoff has a rooted aversion to the Persians which he has taken no pains to conceal; his conduct has been far from conciliatory, and calculated to increase the animosity between the two nations which is already too strong",[19] It should be said, however, that the most intelligent English observers of Persian affairs about this time held much the same views regarding the Persians as did Griboedov. James Morier gave immortal expression to his amused contempt in *Hajji Baba.* John Macdonald, Willock's successor, has left the following pen-sketches of the leading figures in Fath Ali Shah's government:

"The Azeem-oo-Dowleh, though sufficiently well versed in the customary modes of raising money, collecting the revenue, and other little details of internal economy, yet knows nothing of higher matters, or of the foreign policy of the Empire. Meerza Abdool Wahab, the Moatumud-oo-Dowleh and principal adviser of His Majesty in great Affairs of State, is a poor infirm slumbering dotard, tainted and tinctured by a spirit of bigotry, proud of his descent from the Royal House of Sefi, a Syed by birth, a secret enemy of the Kujer family, and one of those intolerant Mohamedans, who deem it pollution to hold communion with infidels. And Meerza Abdool Hussan Khan, the Minister for Foreign Affairs, is a low-minded corrupt and blustering personage, upheld at Court by our support, and possessed of no further weight, or influence, than our countenance affords him. Your Lordship will, I am sure, agree with me, that an administration composed of such materials, is ill suited to guide the Vessel of the State through the storms and tempests which assail it on every side."[20]

One could quote other documents to show that, in experiencing contempt for the Persian officials with whom he came into contact, Griboedov was at one with his British opposite numbers. I would add that in more than one place in his correspondence and his fragmentary diaries he reveals a certain sympathy for the wretched common people of Persia; moreover, that, by his diligent study of the Persian language and literature, he showed that he respected the culture of the country. In a letter (21 June 1824) reporting a conversation with Krylov, Vjazemskij writes: "[Krylov maintained] that poetry should have a *but,* that you cannot attach the body of a bird to the head of a beautiful woman. Oh yes, you can, respected Ivan Andreevič, and this may give rise to a beautiful ideal

[18] *Polnoe sobranie,* III, 143.
[19] Foreign Office Records, 60/14.
[20] Foreign Office Records, 60/31.

Nature, far above that which is seen by our eyes. Have you ever heard of the Griffin of Indo-Bactrian origin? Look at it in the ruins of Persepolis, in the poem of Ferdousi — ah!"[21] It was rather bold in 1824 to refute Krylov — and Horace — by reference to the practice of Persian literature and art. I draw attention to this in order to show that Griboedov's "anti-Persian" feeling did not arise from the vulgar prejudice of a European for Asiatics but from the exasperation inevitably engendered by the "delays, evasions, thriftless lies and miserably shifts of the authorities at Tahraun".[22]

To return to Griboedov's first spell of diplomatic service in Persia, which lasted until November 1821, the extant documents throw light upon two further incidents, one trifling, the other serious, both of them characteristic:

The first was a clash in August 1820 with the British Chargé d'Affaires over the alleged encouragement of desertion among the Russian troops on the Persian frontier by an English officer returning from a visit to Georgia.[23] Griboedov's charge was vigorously rebutted by Willock. At this distance it is hard to be sure who was in the right; my impression is that Willock was. What is interesting is the impetuosity with which now, as in his negotiations with Prince ʿAbbas the year before, Griboedov came forward as champion of the rights and dignity of his country.

The other matter was concerned with the outbreak of the Perso-Turkish war of 1821-1823, for which Griboedov later claimed the "credit". The contemporary evidence for this is rather thin. On 3 December 1828, at a time when Russia was at war with Turkey and Persian assistance might have been useful, Griboedov, now Russian Minister to Persia, wrote to the new C. in C. in the Caucasus, Paskevič: "Stir up our drowsy Ministry of External and Extraordinary[24] Affairs. Write direct to the Emperor, giving your opinion regarding ʿAbbas Mirza, namely, that it would be a good thing to arm him against the Turks. I can't take this upon myself without special authority from my chiefs. In 1821 I did precisely this very successfully and got a wigging from Nesselrode, although Ermolov fully approved me."[25] Another Russian authority for Griboedov's part in the outbreak of the Perso-Turkish war of 1821-1823 is N. N. Murav'ev-Karskij, who in his Memoirs (entry of 2 Feb. 1822) wrote: "Griboedov came to dinner with me. ... He told me, among other things, about Mazarovič's behaviour in Persia and the way his conduct lowers the dignity of his post and, consequently, of our Emperor. When despatches were received from Petersburg informing the Persian court that the Turks were incurring the displeasure of the

[21] *Literaturnoe nasledstvo*, XLVII-XLVIII (Moscow, 1946), p. 229.
[22] Macdonald's despatch of 18 January 1828. India Office Records, Persia, 42 (1828).
[23] D. P. Costello, "Griboyedov in Persia in 1820", in *Oxford Slavonic Papers*, Vol. V (1954).
[24] In the original иностранных и престранных.
[25] *Polnoe sobranie*, III, 239.

Emperor by their conduct and provoking us to war, ʿAbbas Mirza had to be informed that the Emperor wished all nations to know that he was not moved by any desire for conquest but merely by the improper actions of the Turks in his regard. Griboedov went to the Mirza and explained this to him, saying that the Emperor was not asking for allies but was merely informing him of the fact. ʿAbbas Mirza, delighted, promised to put 50,000 men in the field and march against the Turks, as in fact he did. The next day Mazarovič, meeting him in a garden, broached the same subject with him; but, instead of preserving a proper decorum, he requested ʿAbbas Mirza to be our ally, and when ʿAbbas expressed his consent, seized his hand and kissed it in token of gratitude. There you have a gesture worthy of a foreigner, a hireling in our service."[26]

Griboedov's action in encouraging ʿAbbas to go to war with Turkey in 1821 appears curious, to say the least. In spite of what Murav'ev says in the passage just quoted, Griboedov and Mazarovič did not receive any instructions in this sense from St. Petersburg; the "wigging" which Griboedov says in his letter to Paskevič he received from Nesselrode makes this clear. S. V. Šostakovič, the author of the most recent study of Griboedov's diplomatic activity,[27] in fact states that the only authority which they had was a letter of 1 July from Ermolov requesting the Russian Chargé to "devise means of averting" a military alliance between Persia and Turkey. There may have been, in addition, some private communication from Ermolov which has not survived.[28] At all events, Mazaro-

[26] *A. S. Griboedov, ego žizn' i gibel', v memuarax sovremennikov* (Leningrad, 1929), pp. 115ff.

[27] *Diplomatičeskaja dejatel'nost' A. S. Griboedova* (Moscow 1960), pp. 69 f.

[28] The records in the British Public Records office throw some light on this episode.

In a despatch to the Governor General of India, dated 24 September 1821 from Tehran, Willock spoke of the possibility that Russia might incite Persia to attack Turkey (F.O. 60/40). In a despatch of 19 October to the Foreign Minister, Castlereagh, he reported that the Prince-Governor of Kermanshah, Mohammed ʿAli, had invaded the Turkish Pashalik of Baghdad, and went on:

"I trust no hostilities have taken place on the other frontier [that between Azarbaijan and Turkey — D.P.C.] The Kaem Mukam most solemnly assured me that Persia would never co-operate with Russia against Turkey; but some days after my departure from Tabriz M. Mazarowietz received dispatches from his Government, and was earnest in his solicitations with the Prince Royal to commence hostilities against the Turks."

"M. Mazarowietz with a view to making H.R.H. suppose that Great Britain had become a party in the dismemberment of Turkey, informed H.R.H. that Alexandria had been taken by the British Troops. It is to be hoped that H.R.H. has not been led away by this deception, and that he has remained firm to his purpose of overlooking for the present his supposed grievances against the Turks" (*ibid.*).

In his next despatch to Castlereagh, dated 15 December 1821, Willock reported Prince 'Abbas' invasion of Turkey and capture of Bayazid. He continued:

"The Prince Royal, conscious no doubt that his cause required justification, induced M. Mazarowietz to vindicate to the Court the hostilities he meditated against the Turks. This document which was delivered to me by Mirza Abdul Wahab confirms the reports which had previously reached me of the expediency of the measures having been strongly urged by M. Mazarowietz."

vič and Griboedov, in encouraging ʿAbbas to go to war with Turkey, were carrying out the policy, not of their Government, but of the Commander-in-Chief of the Caucasus. The consequence was that, when ʿAbbas did attack Turkey, Russia did not support him. The two years of war, in which Persia squandered considerable resources, both human and material, ended in a stalemate. It was natural that ʿAbbas should consider that he had been let down by Russia and should harbour resentment against the diplomats who had misled him. Griboedov showed that he had drawn the lessons of this episode when

---

This remarkable document, by which the head of the Russian Legation took it upon himself to justify to the Persian Government the action of the Persian Crown Prince in invading a neighbouring country without instructions to this effect from the Shah, is preserved in translation in the Foreign Office Records. It is addressed from Tabriz to the Chief Eunuch, Agha Manuchehr Khan, and is dated 29 August O.S. [1821].

"I have been honoured by two Firmans from His Persian Majesty. Shameer Khan, Interpreter, has informed me of the honours conferred upon him at Court, and I have gratefully to acknowledge these acceptable proofs of His Majesty's favour. General Yermoloff has learnt the gracious reception of his presents, and the Rukum addressed by His Majesty to His Excellency has been forwarded.

"I am happy to learn that the Chiefs of Khorasan have tendered their submission, and have sent hostages in pledge of their future services, and that H. R. H. Hussun Ali Mirza has defeated Bunead Beg, and restored the country he governs to a state of tranquillity. The success obtained by Mahomed Kooli Mirza over the Turcomans is also satisfactory.

"During the period of my residence in Persia the sons of His Majesty have continually overcome their enemies; the good fortune which has aided their arms must be ascribed to the justice of the causes they have espoused, and to the energetic and liberal spirit which has marked their administration. In like manner the Prince Royal has for a long time patiently endured the unseemly and unfriendly acts of the Turkish authorities, and has endeavoured by amity and conciliation to produce a more pacific disposition on their part. The mild measures of H.R.H. have not however been attended with beneficial effect; on the contrary the conduct of the Turks has assumed a more outrageous and overbearing aspect.

"Hafiz Ali Pacha Sirasker of Erzeeroom has scandalously illtreated the Persian pilgrims and Merchants.

"The Pacha of Moosh seduced from their allegiance to Persia the Hyderanloo and Sebekee Curds, and regardless of the obligations resulting from the amity of the two States afforded them protection. Hussun Khan who marched into Turkey for the purpose of bringing back these Tribes, and considered himself in the Territory of a friendly State was attacked by the troops of the Pacha of Moosh, and was placed under the necessity of defending himself. Selim Pacha derived no advantage from this attack, but the object is to shew that those who disturb the friendly relations of the two States are honoured and distinguished by the Turkish Government. The Pacha in reward for the opposition offered to Hussun Khan was created Sir Asker.

"Derweesh Pacha rebelled against his Government, and sought the protection of the Prince Royal. The good faith of H.R.H. rejected his insidious proposals and the Pacha hopeless and dejected returned to the Turkish dominions, and was reinstated in the Government of Van at the general call of the people of that Province. Why did Hafiz Ali Pacha conceal these facts from his Government?

"Mahmood Pacha the Governor of Van killed an officer of the Beglerbeggee of Khoi who demanded payment for grain sold to the Turks by Persian subjects.

"Persian travellers and merchants experience continual ill treatment throughout the Turkish dominions.

"Saudik Pacha fled from Turkey, and sought refuge in Persia. H.R.H. Abbas Mirza sent him he wrote, in the letter to Paskevič of 3 December 1828 (quoted above), "'Abbas Mirza won't believe me [i.e., when encouraged to go to war with the Turks — D.P.C.] unless I categorically declare to him the will of the Emperor."

He elaborates the same theme in a letter to Bulgarin of the end of November 1828: "You keep insisting in your letters that I get 'Abbas Mirza to go to war with the Turks. My dear fellow, do you know whether I have permission to do so? I am surprised that someone who knows the answer to this better than you do should say that I could do this. If you are in the service of the State then the

---

to Turkey accompanied by a Mehmandar and requested his pardon. The Pacha of Erzeeroom imprisoned the Persian officer and Saudik Pacha during four months, and afterwards liberated the former without offering any apology or explanation to H.R.H., and to shew his respect for the intercession of the Prince cut off the heads of Saudik Pacha and of twenty of his followers in the market place, and sent them to Constantinople.

"Hajee Ali Beg a respectable person who was charged by the Prince Royal with communications to the Sir Asker has now been confined eight or nine months at Erzeeroom.

"Arms which were purchased in France for the use of the Persian Troops, have been detained without any reasonable cause being assigned at Constantinople.

"These continued injuries have exasperated beyond measure the Prince Royal, and H.R.H. solicitous to preserve unblemished the National honour of Persia, having made the necessary preparations, has marched to punish the insolence of these arrogant people.

"My hope is in the assistance of God; and the effective measures which H.R.H. has adopted give me every confidence that the object he has in view will be attained.

"I trust his Persian Majesty will be equally anxious to maintain the National dignity, and will approve of the measures pursued by His Royal Highness.

"Although I ought to have addressed Mirza Abdul Wahab on this subject, yet as I have lately written a letter of condolence to His Excellency, I have not troubled him on the present occasion' [Mirza Abdul Kerim, son of Abdul Wahab, had died under suspicious circumstances shortly after breakfasting with Allah-Yar Khan, who was believed to have poisoned him — Willock's despatch no. 14 of 18 October 1821].

"I trust you will find a convenient opportunity of explaining to His Persian Majesty the contents of this letter, and I have to offer Your Excellency the assurance of my high respect and consideration.

sealed — Simon Mazarowietz
a true translation — Henry Willock".
(Foreign Office Records, 60, Persia, 1822).

In a report which Willock addressed from Tabriz to Lord Strangford, the British Ambassador at Constantinople, on 30 April 1822, he indicated that Persia was now ready to make peace with Turkey:

"The prospect of peace between Russia and Turkey which is now presented makes H.R.H. suppose that the power of the Ottoman Empire may be exerted in the recovery of her lost provinces and H.R.H., who would not have undertaken the invasion of Turkey last autumn had he not been informed that no opposition could be offered to his troops, is not solicitous of engaging himself in a warfare both unprofitable and doubtful in its result.

"The counsel of Monsieur Mazarowitch is also at this moment pacific, and from an expression of His Royal Highness in a letter lately addressed to the Emperor of Russia which states 'that the paternal advice of His Imperial Majesty shall be attended to' it may be inferred that the Emperor considered the hostilities of last autumn premature." (*ibid.*).

Prince 'Abbas' letter to the Emperor Alexander is quoted at greater length in Willock's despatch of 4 May to Canning:

thing you have to do is, above all, to follow literally the instructions you receive from above. If at the same time you can also manage to do something that gets into the newspapers, then so much the better. I'm prepared to think up all sorts of offensive plans, but not to put them into action until I am instructed to proceed thus and not otherwise."[29]

I would draw attention to what may have been another reference to this divergence of policy between St. Petersburg and Tiflis in 1821. When, in 1827, during the Russo-Persian war, Griboedov was engaged in peace discussions with Prince ʿAbbas, in the course of the argument he contrasted the state of affairs in Russia, where one will, that of the Emperor, was absolute, with that in Persia, where the authority of the Shah was constantly eroded by the actions of his subordinates. To this ʿAbbas replied: "There is no single will in Russia either: in Petersburg they say one thing, Ermolov says something else."[30] At the time of speaking Ermolov was no longer in command. ʿAbbas, therefore, was speaking of the past, possibly of the events of 1821. Such a veiled reference to a matter that closely concerned the enemy negotiator would have been true to the style of Persian diplomacy.

The blame for the gaffe which Mazarovič and Griboedov made in 1821 does not lie on them but upon Ermolov, from whom they received their immediate instructions. What is odd is that Šostakovič[31] should represent what in effect was a provocation as a master-stroke of diplomacy.

In November 1821 Griboedov was sent to Tiflis to report to Ermolov on the situation in Persia. On the way he broke his arm in two places and used this mishap as grounds for a request to be transferred from service in Persia. He

---

"H.R.H. in a letter to the Emperor Alexander says, I have received your paternal advice which shall be attended to. It is for Y.I.M. to suggest and for me to obey. My late conduct to Turkey was adopted in pursuit of Y.I.M.'s wishes and I shall still look to them for my future guidance" (*ibid.*).

Finally, in a letter dated 6 and 8 August 1822, George Willock, Chargé d'affaires *ad interim*, wrote to his brother Henry, who was now back in London:

"M. Mazarowietz has I think been reprimanded by his Government for his forwardness in promoting the Turkish war. He has complained to me that while you testified to him every kindness, your representations to the Government were unfavourable, and that you wrote to Lord Strangford that he was the chief cause of the war. I fancy Lord Strangford must have had some communication with St. Petersburgh on this subject, as M. Mazarowietz has either received extracts from your despatches, or has been reprimanded on account of their contents.

"I do not entirely comprehend the business, but General Yermoloff explicitly told Mirza Sauleh [Prince ʿAbbas' Private Secretary, who in the summer of 1822 journeyed via Tiflis and St. Petersburg to London to transact business with the British Government — D.P.C.] that you reported to Lord Strangford that M. Mazarowietz was most active in widening the breach between the two countries" (*ibid.*).

[29] *Polnoe sobranie*, III, 232.

[30] *Op. cit.*, p. 257.

[31] *Op. cit.*, pp. 66-71.

was appointed Diplomatic Secretary to Ermolov and in that capacity remained in Russian Transcaucasia until February 1823, when he was given four months' leave to return to European Russia. He contrived to have this extended to well over two years and did not return to the Caucasus until the end of September 1825. This spell in Russia proper was the period of the completion of *Gore ot uma.* As I am concerned here not with his literary achievement but with his diplomatic activities, I pass on to consider the further course of his career.

In January 1826 Griboedov was arrested at Groznaja in the Caucasus on suspicion of involvement in the Decembrist conspiracy and was conducted under escort to St. Petersburg. There he was held under arrest for something like four months, but was finally cleared of the charges against him and released in June 1826. He was given his travelling expenses for the return to Georgia as well as promotion to the rank of *nadvornyj sovetnik.* Dawdling as usual when on his way to report for duty, he reached Tiflis only on 3 September.

In the meantime war had broken out with Persia. Ermolov's excessive caution in the opening phase displeased the Tsar, and in March 1827 he was superseded by Paskevič. The new Commander-in-Chief was married to Griboedov's first cousin, and Griboedov was naturally maintained in the post of Diplomatic Secretary.

The following months, until the conclusion in February 1828 of the Treaty of Turkmanchai, were the period of his most brilliantly successful diplomatic activities. On three occasions he took a leading part in peace negotiations with the Persians. The first time, at Karaziadin in July, he was the sole Russian negotiator; his chief interlocutor was Prince ʿAbbas himself. Griboedov has left an account of these discussions,[32] in which the Persians, always evasive, are represented as indulging in what Griboedov terms "the dialectic of the thirteenth century," while he patiently and firmly insisted on the conditions which his commander had prescribed. The conversations led to nothing: ʿAbbas evaded precise obligations and asked for a long armistice, which it was not in Russia's interest to grant him. At the end of his report to Paskevič on these negotiations, Griboedov expressed the opinion that the Persians could not be brought to accept Russia's demands (the cession of the provinces of Erivan and Nakhjavan and the payment of an indemnity) until the Russians should have captured Erivan — "a highly important fortress (the Bhurtpore of Persia) and on which the fate of the campaign will depend."[33]

Erivan did, in fact, fall on 1 October; and on the 13th the Russians occupied ʿAbbas' capital, Tabriz. Nine days later ʿAbbas' vizier, the Qaem Maqam, repaired to the Russian camp, stating that he had full powers to negotiate for

[32] *Polnoe sobranie*, III, 252-267.

[33] Macdonald to the Secret Committee of the Court of Directors, 21 April 1827, F.O. 60/30.

peace. The two parties agreed that Persia should cede Erivan, Nakhjavan and part of Talish and pay an indemnity of 15 crores (7½ millions) of tomans (1 toman = between 10 and 12 shillings or between 2½ and 3 dollars): 5 crores to be paid in one month, the other ten in three months. If the ten crores were not paid within three months the original five would be forfeited and the province of Azarbaijan would be ceded to Russia. ʿAbbas accepted these terms, leaving the remaining articles of the treaty to be settled at a later conference at Dei Kargan, where the plenipotentiaries met on 6 November. Griboedov was again the chief Russian negotiator. In consideration of a request from ʿAbbas, the total sum claimed by the Russians was reduced from 15 to 10 crores, on condition the first 5 were produced on 22 November. The Persians were subsequently granted an extension of time until 10 December, but this date too passed without payment being made. At this point the Shah suddenly revoked his son's powers and produced an entirely new set of proposals, which Paskevič found quite unacceptable. In great irritation he broke off negotiations and prepared to march on Tehran.

At last, thanks largely to the pressure applied by Macdonald, the Persians agreed to pay. In a letter to the Secretary of the Government at Fort William (Calcutta) Macdonald explained that he had intervened only in order to save Persia from dismemberment. Paskevič's forces, he said, were "more than sufficient in the present urgency to subvert the Persian Monarchy." He went on: "Nothing indeed short of a positive conviction of this fact, combined with a dread of the unpleasant consequences likely to accrue to us from the overthrow of the reigning family could have induced me to press with so much urgency the payment of the first instalment as the only means of obtaining the signature of the Treaty."[34]

The peace between Russia and Persia was signed at the village of Turkmanchai on 10 February 1828. Macdonald described the occasion in his Private Journal: "The Minister for Foreign Affairs and the Kaim Mukam have been true to their pledge. Abbas Meerza has also kept his promise, and in the dead of night a Royal Salute of one hundred guns announced to the Inhabitants of Toorkoomanchaee, that Persia had again become the ally of Russia. The whole of the morning was passed in the mutual shaking of hands, and in the interchange of congratulations. The Prince held a Levee in a chamber about six feet square, and the General gave a grand dinner in a barn, so dark that candles were required many hours before the setting of the sun. The healths of the Emperor, George the 4th and of the Shah were drunk in champagne with great applause. His Excellency appeared to be in great pain, having while on a visit to the Assuf-

[34] India Office Records, Persia 42.

oo-Dowleh fallen into a tunnoor (oven), and inflicted a scald as well as a dangerous wound in his leg."[35]

Severe as the Russian terms were, Macdonald nevertheless considered that the conclusion of peace was advantageous both to Persia and to England, since without it the Russians might have put an end to the existence of the Persian State. "To seize on Gilan and Mazenderan, to revolutionize the Government, and to dismember the Empire into a number of petty principalities, or khanats, as they are styled, was, I have just reason to imagine, the intention of the Russians, had their army once crossed the Kizel-OOzun, nor would it have been difficult, in the present unsettled state of mens' minds, to have carried that plan into immediate effect."[36] That Macdonald was right, at least in part, in his interpretation of Russian intentions is shown by the fact that in 1827 Griboedov composed a report on the Persian province of Gilan,[37] in which he drew attention to its wealth in silk and timber, to its geographical isolation from the rest of Persia and to the disaffection of its inhabitants to the reigning dynasty.[38] It would be easy, he suggested, to detach this province from the rest of Persia.

Griboedov had played the leading part in the negotiations both at Die Kargan and at Turkmanchai. Murav'ev-Karskij, who knew him well, and who disliked him personally, wrote that without Griboedov's help Paskevič would not, perhaps, have been able to cope with the difficult situation in Georgia in 1826 and 1827 and "could not have concluded such an advantageous peace with Persia."[39]

Paskevič himself in his report to the Tsar wrote of Griboedov: "I venture to recommend him as a man who has been extremely useful to me on the political side. I am indebted to him for the idea of not proceeding to the conclusion of the treaty without previous receipt of part of the money; and the event has shown that, had it not been for this, we should not for a long time have achieved in this matter the success desired."[40]

From a man as vain as Paskevič this was high praise.

Griboedov was sent to St. Petersburg to take the text of the treaty for ratification by the Tsar. He was received most graciously, was given every mark of distinction and, in April, was appointed Minister Plenipotentiary to Persia. His

[35] India Office Records, Persia 43.

[36] Letter to George Swinton, etc. etc. of 22 February 1828. India Office Records, Persia 42.

[37] *Polnoe sobranie*, III, 249 ff.

[38] That Griboedov's information on this point was correct is indicated by the following extract from Henry Willock's despatch to Castlereagh, dated 10 February 1821: "The inhabitants of Gilan have loudly complained of the tyranny of their Governor the Prince Mahomed Reza Mirza; it is even reported, and generally credited, that they sent Deputies offering submission to the Russian Government" (Foreign Office Records, 60/20).

[39] *Op. cit.*, p. 138.

[40] *Polnoe sobranie*, I, lxxix.

duties in his new post, apart from the normal functions of a head of mission, included the superintendance of the execution of the peace treaty terms, the most troublesome of which, given the ruinous condition of Persia's finances and the notorious miserliness of the Shah, was going to be the collection of the outstanding crore of tomans.

As usual, Griboedov was in no hurry to get to his post. He arrived in Tiflis on 5 July and did not reach Tabriz until 7 October. This time his tardiness had a sound political motive: he did not wish to arrive in Persia before the indemnity had been paid in full, for fear lest ʿAbbas should think him authorised to waive the payment of the eighth crore. For this reason he considered that it would be politic to wait in Tiflis.[41] However, on the imperative instructions of the Ministry in St. Petersburg and against his better judgement, he left Georgia and arrived in Persia before the last crore had been paid in full. On 30 October 1828 he wrote from Tabriz to Rodofinikin, the head of the Asiatic Department of the Ministry of Foreign Affairs:

"From the point of view of the tasks assigned to me I got here too soon. I knew this in advance, but was afraid to have to answer to the authorities, who in our country measure success and zeal in the execution of tasks by reference to the greater or less speed at which officials travel. From Tiflis I could have threatened ʿAbbas Mirza that I would not come at all if he did not pay in full the money he owed us. This was the line I took to begin with, and he, knowing that my arrival was the guarantee of his security and of Russian protection, hastened to begin payment. If I had waited for another month before coming here and had continued my manoeuvre, the whole of the 8th crore would by now have been in our hands. As things are, what has happened? I am pestered from morning until late at night with absurd proposals, I am urgently requested to let them off, first 200,000, then, 100,000 then 50,000 tomans. Their arguments are irrefutable: they are utterly ruined. I, of course, don't agree to anything they propose. But things do not get any further ahead. Before my arrival here 200,000 was extracted from them, before I reached Erivan, and 100,000 since then. But once they heard that I was in Nakhjavan, they refused outright to pay any more. Such is the outlook of the people and government of this country: they greet every newly arrived diplomatic agent as someone invested with far-reaching powers, whose function it is to make them all sorts of concessions, dispensations, gifts, and so forth. In the course of twenty-five days I managed with the utmost difficulty to squeeze 50,000 out of them over and above the 300,000. To get the remaining 150,000 I am going to Tehran, where ʿAbbas Mirza has sent McNeill to negotiate a loan of 100,000 from his father; I

[41] Letter of mid-August 1828 to A. K. Amburger, *Polnoe sobranie*, III, 223.

shall have to reinforce McNeill's actions by putting pressure upon the Shah."[42]

By the beginning of December 1828 a further 50,000 tomans had been paid, leaving 100,000 outstanding.

A few days before Christmas Griboedov left Tabriz for Tehran for the purpose of presenting his credentials and of settling a few outstanding problems (neither of the two main sources for the events of his stay in Tehran, I. S. Malcev and the author of the "Narrative of the Proceedings of the Russian Mission",[43] mentions the payment of the 100,000 tomans). He reached Tehran about the New Year.

It is not my purpose here to trace in detail the course of events which led on 30 January to the assassination of Griboedov and practically the entire staff of the Russian Legation. The essentials of the story have already been told by N. K. Piksanov in the biographical sketch prefixed to the first volume of the Academic edition of Griboedov's works. Further details are to be found in Dr. D. M. Lang's paper "Griboedov's Last Years in Persia",[44] and in mine "The Murder of Griboyedov".[45]

The facts, in outline, are these: at the beginning Griboedov's mission was eminently successful; the Shah granted him all his requests and lavished honours on him and his staff. It was only when Griboedov, his business in Tehran completed, was preparing to return to Tabriz that the crisis suddenly broke and developed quickly to a climax. The chief eunuch of the Shah's harem, Mirza Ya'qub, availing himself of a clause in the Treaty of Turkmanchai, took refuge in the Russian Legation and asked to be repatriated to his native Armenia, now Russian territory. Griboedov endeavoured to persuade him to change his mind, but without success. He accordingly kept the eunuch in the Legation premises, intending to take him with him when he returned in a few days' time to the north. This was the chance for which Griboedov's enemies had been looking since the Shah was determined at all costs to prevent one who knew the secrets of his sentimental life from leaving the country. The Shah's concern in the matter was purely personal. On the other hand, some of the most powerful men in Persia, including the Shah's son-in-law, Allah-Yar Khan, and in particular the rivals and enemies of Prince 'Abbas, who looked on him as a Russian protégé, were pursuing political aims. Feeling that they stood to lose nothing and possibly to gain much from a rupture with Russia, they were anxious to push matters to an extremity, and the Shah's rage at the defection of Mirza Ya'qub made it possible for them to do so. With the cooperation of some fanatic or venal mullahs they roused the city rabble and instigated an assault in

[42] *Op. cit.*, p. 228.
[43] *Blackwood's Magazine*, September, 1830.
[44] *American Slavic and East European Review*, VII, No. 4 (December 1948), 317-339.
[45] *Oxford Slavonic Papers*, VIII (1958), 66-89.

overwhelming strength on the Russian legation. The authorities refrained from intervention until the massacre was over.

It is noteworthy that the cause in which Griboedov lost his life was essentially that which he had defended so passionately against Prince ʿAbbas in Tabriz nearly ten years before — the right of Russian subjects to return to their own country.

I cannot leave this subject without saying something about the endeavours of certain Russian writers to prove that the murder of Griboedov and his staff was organized by the British. I have shown[46] that this was not the case, and I shall not repeat my arguments here, except to say that the guiding principle of British policy in Persia at this time was to prevent anything that might have given the Russians an excuse for intervention in Persia. I would, however, mention one point which I have not discussed elsewhere. Šostakovič[47] makes a good deal of the Russophobia of Lord Ellenborough, who in 1828 became President of the Board of the East India Company, and suggests that the upheaval in which Griboedov lost his life was organized on Ellenborough's instructions. To substantiate the charge, he quotes Mr. C. H. Phillips who says:[48] "He [Ellenborough] hoped to counteract Russian influence in Persia by organising disturbances against Russia in that country." Šostakovič does *not* give the rest of the sentence, which runs: "but Wellington[49] was not disposed thus to alienate Russia." If the Prime Minister was opposed to Ellenborough's policy one can be sure that it was not applied; the situation in Britain was not like that in Russia in the days when Nesselrode and Ermolov pursued each a policy of his own.

Šostakovič plays the same trick with a quotation from *Ellenborough's Political Diary 1828-30*. Ellenborough wrote in his diary on 10 October 1828: "I would, in Persia and everywhere, endeavour to create the means of throwing the whole world in arms upon Russia at the first convenient time." Mr. Šostakovič quotes this but does *not* quote, form the same paragraph: "The Duke ... is for the more pacific line, and would have the Persians give their great neighbour no justifiable ground for interference with them."

If Šostakovič was so firmly convinced of Lord Ellenborough's responsibility for the crime of Tehran one might have expected him to look up the passage in the *Political Diary* in which Ellenborough expresses his reaction to the news of it. Perhaps he did; if so, he must have been disappointed; for, instead of the homicidal glee which Šostakovič's arguments might have led one to expect,

[46] *Ibid.*
[47] *Op. cit.*, pp. 227 f.
[48] *The East India Company 1784-1834*, p. 269.
[49] The British Prime Minister.

Ellenborough expresses regret — a regret deriving entirely from political considerations. This is the entry, dated 14 April 1829:

"By the newspapers I see that there has been a quarrel at Tehran between some of the Russian Ambassador's suite and the populace, which led to an attack on the Russian palace, and to the death of the Ambassador and all his people except two. This is an unfortunate event, as it will give the Russians a new claim to indemnity, which they will exercise inexorably. Probably they will insist on the junction of Persia in the attack on Turkey, as the only satisfaction they can accept. It is just possible that the example once given, and the people despairing of pardon, a rising against the Russians may take place, and something of a national feeling arise in Persia. But I fear this will not be the case. I suppose our Minister was at Tabriz."[50]

Ellenborough, therefore, had nothing to do with the disorders in Tehran.

The treatment by Šostakovič of the question of Lord Ellenborough's possible connection with the Tehran massacre illustrates the spirit in which some (by no means all) Soviet historians address themselves to such subjects as that of Griboedov's diplomatic activity. The fact that they do not stick at suppressing any documents that contradict their thesis demonstrates that they do not themselves believe a word of what they say. They approach their theme, not as scholars hopeful of attaining to the truth, but as lawyers for whom any argument will do, so long as it seems to support their case. For those who relate what in fact happened they have the usual battery of tired abuse, of which "bourgeois apologist" and "colonialist falsifier" are fair samples. (Expressions taken from a review of Šostakovič's book by V. T. Pašuto.[51] Pašuto, who has himself written on Griboedov's diplomatic activity,[52] is, like Mr. Šostakovič, a representative of the old-established Špekin[53] school of Russian historiography).

It remains to attempt a brief appreciation of Griboedov's qualities as a diplomat.

Wherever there was a diplomatic task which required intelligence, firmness, and courage, there was no one in Russia better qualified than Griboedov to perform it. Add to this his thorough knowledge of the Persian language, which is well attested, and his interest in Persian history and literature, and it is obvious that Nicholas I could not have made a better choice than Griboedov for the post of Russian Minister in Persia. One cannot but accept the judgement of Murav'ev-Karskij:

[50] Lord Ellenborough, *A Political Diary 1828-30* (London, 1881), II, 15.
[51] *Istorija SSSR*, No. 4, 1961, pp. 184-186.
[52] In *Istoričeskie zapiski*, XXIV, (1947), 111-159.
[53] Почтмейстер (Шпекин): Это все француз гадит (*Revizor*, Act I, Sc. II).

"While under no misapprehension on the score of Griboedov's much praised virtues and principles, which I never found attractive, I always did full justice to his ability, and I remain convinced that Griboedov in Persia was in absolutely the right place, that by his personality alone he was as useful to us as an army of twenty thousand men, and that a man as well qualified to occupy that post could not, perhaps, be found in Russia"[54] (so much for Pašuto's allegation in the review referred to above, that "the Tsar's government intentionally sent A. S. Griboedov, whom they distrusted because of his Decembrist connections, as far as possible from the capital").

Griboedov, then, was a diplomat of very high quality. Yet from the beginning he was unhappy in the diplomatic service and looked forward to quitting it when a suitable occasion might offer. This dissatisfaction comes out again and again in his correspondence. On 16 April 1827 he wrote to Bulgarin: "Don't expect poetry from me: all my attention is absorbed by mountaineers, Persians, Turks, administrative business, and the vast correspondence of my present chief. Not for long, of course; once the campaign is over I'll bow myself out. I'm of no use in ordinary times, and it's not my fault. People are petty; their affairs are stupid; one's heart grows hard, one's mind becomes fogged, and one's moral sense is destroyed, without any benefit to others. I was born for a different sort of career."[55]

In October 1827 he wrote to a friend: "[I am] determined to go away or to get right out of the service, which I loathe, whatever sort of future it may offer me".[56]

In July 1828, after his appointment to the post of Russian Minister he wrote: "this won't last long and soon, in two years at the most, I shall become the hermit of Tsinandali" (his future wife's family estate).[57]

On the eve of his departure to Tehran, in December 1828, he wrote to Paskevič: "Now all that's needed is backing from the Ministry. If it isn't forthcoming, if they let me down, then goodbye ambition and all that it entails; that is not my ruling passion. Tsinandali and Kakhetia are worth still more, even though I have nothing in the world, apart from my prospects of promotion and a modest pension that would be due to me in time."[58]

One could quote other passages from the correspondence to show that he did not feel at home in the diplomatic service and did not consider himself committed to it for life. The reason was not, as some might be inclined to think, that Griboedov was too liberal in his outlook to serve willingly as the diplomatic agent of the autocratic Nicholas. In fact, he was not by any means as radical in

[54] *Op. cit.*, p. 139.
[55] *Polnoe sobranie*, III, 199.
[56] *Op. cit.*, p. 205.
[57] *Op. cit.*, p. 221.
[58] *Op. cit.*, p. 240.

his views as some Soviet writers make out,[59] and we happen to know that he regarded Nicholas as an excellent master to serve. He wrote to his subordinate, A. K. Amburger, in August 1828: "the Emperor Nicholas has an energy which his predecessors did not possess, and it is pleasant to be the Minister of a Sovereign who sees that the dignity of his envoys is respected."[60]

The reason for his discontent must be sought elsewhere. His judgement on himself — "I'm of no use in ordinary times" — was, I believe, sound. Griboedov's unusual combination of high intelligence with inflexible character made him a first-class representative of his country in time of crisis. This he showed as early as 1819 and later demonstrated on a much bigger stage in 1827-29. But diplomatic life had no interest for him in quiet times. He was too restless and too independent for a placid round of duties and functions, and his letters written during the quiet spells of his career show him plunged in boredom and gloom. He was

> A daring pilot in extremity;
> Pleased with the danger, when the waves went high
> He sought the storms; but for a calm unfit.....

If he could have enjoyed freedom of action, if he could himself have worked out the policies which he applied, he would perhaps have been happy as a diplomat. His fantastic scheme for the establishment of a Caucasian version of the East India Company with himself in command[61] was, I believe, an attempt to create for himself the situation of virtual independence which he needed. Nothing came of this. What would have been his future career if he had not perished in Tehran is an attractive (if vain) subject of speculation. If Nicholas had decided to accept the risk of serious trouble with Britain and to push on vigorously with the policy of reducing Persia to the status of a Russian protectorate, then Griboedov would have been the ideal agent of this policy, which was, we know, his own.[62] But, as Pašuto reproachfully (and revealingly) observes: "Tsarism came out in favour of the integrity of the empire of the Qajars and the maintenance of the existing relations with England."[63] So that Griboedov would not have found the necessary outlet for his energies. He might have retired from the

[59] See "*Griboedev i krepostnoe pravo*", in N. K. Piksanov, *Griboedov*, pp. 146ff.

[60] *Polnoe sobranie*, III, 222.

[61] See O. P. Markova, "Novye materialy o proekte rossijskoj zakavkazskoj kompanii A. S. Griboedova i P. D. Zavelejskogo", in *Istoričeskij arxiv*, VI (1951), 324-390.

[62] See the extract from Willock's despatch of 3 October (N.S.) 1819, quoted on pp. 55-56, and cf. "We should ... solemnly promise that we shall elevate him ['Abbas] to the throne, since this will cost us nothing ... while our influence in Asia will become predominant over that of all other Powers" (letter of Griboedov to Paskevič of 3 December 1828, *Polnoe sobranie*, III, 240).

[63] *Istoria SSSR*, No. 4 (1961), p. 186.

service and settled down somewhere, perhaps in Georgia, as he himself suggested he would in two letters quoted above. Or, alternatively, he might have found that family responsibilities made this course impossible — as he said in the letter in which he informed a friend of his betrothal, "I shall work hard for the Tsar so as to have the wherewithal to feed my children."[64] In that case one can imagine him oppressed by a sense of frustration and failure, growing ever more embittered and sardonic, the terror of his colleagues.

Talleyrand is said to have warned the staff of the French Foreign Ministry, "Above all, no enthusiasm!" Griboedov had too much enthusiasm in him to make a successful career diplomat.

64 *Polnoe sobranie*, III, 221.

THE LIFE OF LADY MOROZOVA AS LITERATURE

HORACE W. DEWEY

University of Michigan

A special place in pre-Petrine Russian literature belongs to seventeenth-century writings by or about Old Believers, particularly the autobiographical *vita* of Avvakum, which is known to many readers and recognized as a masterpiece. From the same epoch comes another work which has somehow received little attention, despite its considerable literary merits. It is the biography of Avvakum's follower, Lady Morozova, written by an unknown but talented author in the 1680's.[1] Millions know the Lady Morozova from Surikov's famous painting, now hanging in Moscow's Tretjakov gallery, which frequently finds its way into collections of color reproductions. Yet the literary account of Lady Morozova's life and death remains strangely neglected.[2]

[1] Some scholars believe the author was Lady Morozova's brother; Pierre Pascal, *Avvakum et les débuts du raskol* (Paris, 1938), p. 338, fn. 74, and P. S. Smirnov, *Vnutrennie voprosy v raskole XVII v.* (St. Petersburg, 1898), p. xci. Our text is taken from N. Subbotin, ed., *Materialy dlja istorii raskola za pervoe vremja ego suščestvovanija*, VIII (Moscow, 1887), pp. 137-203. Although two other women are mentioned in the title (*Žitie bojaryni Morozovoj, knjagini Urusovoj i Mar'i Danilovoj*) the work really centers on Lady Morozova. Subbotin's texts (a variant is presented with the help of extensive footnotes) comprise the so-called "long" or "extended" (*prostrannaja*) version; a later "short" version exists, of which nine copies — dating from 1786 to the early twentieth century — have so far been found, but none have yet been published: A. I. Mazunin, "Kratkaja redakcija Povesti o bojaryni Morozovoj", *Trudy Otdela drevnerusskoj literatury*, XVIII (Moscow-Leningrad, 1962), pp. 341-351. Our paper deals only with the long version.

[2] Of contemporary anthologies, only the late Adolf Stender-Petersen's *Anthology of Old Russian Literature* (New York, 1954) includes the work, and then only in fragments (pp. 387-404). One searches in vain for any reference to the work in most of the standard literary histories (by Čiževskyj, Istrin, Kajev, Orlov, Speranskij, Stender-Petersen [!], Vodovozov and others) or in N. F. Droblenkova's comprehensive *Bibliografija sovetskix russkix rabot po literature XI-XVII vekov* 1917-1957 *gg.* (Moscow-Leningrad, 1961), 434 pp. Three pages are devoted to it by M. O. Skripil' in the Soviet Academy of Science's *Istorija russkoj literatury*, II, Pt. 2 (Moscow-Leningrad, 1948), pp. 329-332. Pre-Soviet studies such as N. S. Tixonravov's chapter on Lady Morozova in *Sočinenija. II. Russkaja literatura XVII i XVIII vv.* (Moscow, 1898), 12-52, treat the work as a historical account and biography. The author of this paper was taken to task by A. I. Mazunin for stating, in 1961, that Soviet scholars had "largely neglected" the work; "Replika amerikanskomu učenomu professoru G. Djui", *Russkaja literatura*, No. 3 (1962), pp. 223-224, but even Mazunin was able to turn up only two earlier

There is something about her story that moves us, even on a first reading. A closer examination of the work reveals the author's mastery — in some cases, perhaps, an unwitting mastery — of certain important features of narrative art. Thanks to his skill, Lady Morozova is not just another stylized religious fanatic; her tale arouses genuine interest.

The author of the *Life of Lady Morozova* pays obeisance to the structural formula which underlies most saints' lives. One variant of the *Life* starts with an extended introduction which surveys events up to the time of Nikon's reforms. It is a flowery, jingoistic passage containing allusions to the Old Testament, the Holy Trinity, Cyril the Philosopher, Vladimir and the Christianization of *Rus'*, the White Cowl, the Imperial Book of Degrees, Moscow the Third Rome, the introduction of printing in Russia and, finally, Nikon's reforms. After the first paragraph the introduction has nothing to do with Lady Morozova, and may have been added later to the original *Life*. Most modern readers will probably consider it the weakest section of the entire work, but such material was often included in saints' lives.[3]

When the author takes up the story of Theodosia (the future Lady Morozova), he follows that part of the *žitie* formula which tells of the saint's family background and childhood years, but he stresses only the high birth and piety of her parents. Nothing is said about Theodosia's early life until her marriage, at the age of seventeen, to the boyar Gleb Ivanovič Morozov. Some sort of miracle (again typical of hagiographical writing) is suggested in connection with the birth of her son, which occurred "after a certain appearance of St. Sergius the Wonder-Worker". In a few years Theodosia became a widow. Shortly after her husband's death she learned of Nikon's "corruption of the church's sacred traditions and rituals and holy writings, and of its spiritual songs and prayers". We are not told just when Theodosia became a follower of Avvakum, but she had obviously done so early in the schism. She shunned Nikon's "innovations" and word of her opposition quickly reached the Tsar.

At this point we note the author's first recourse to a favored structural device — the confrontation — whose repeated use in the *Life* provides the real framework of the story. At least ten confrontations are reported in some detail in the

---

Soviet titles: M. O. Skripil's contribution (mentioned above) and V. A. Desnickij, ed., *Istorija russkoj literatury*, I, Pt. I (Moscow, 1941), p. 304. The latter is practically worthless as a source, devoting a single paragraph of twelve lines to the work. Mazunin himself has published two valuable studies: "Ob odnoj pererabotke žitija bojaryni Morozovoj", *Trudy Otdela drevnerusskoj literatury*, XVII (1961), pp. 429-434, and the 1962 article we have listed in our first footnote.

[3] Avvakum's autobiography has an even more elaborate introduction. A much earlier example may be found in Nestor's *Lection on the Life and Assassination of SS. Boris and Gleb.*

Life, and most are accompanied by dialogue — some of the finest dialogue in Old Russian literature.

The first confrontation is described briefly. The author simply states that the tsar, after himself failing to win Theodosia over to Nikon's reforms, sent Archmandrite Joachim and an associate from the Miracle Monastery to see her. She "strongly resisted them", and, having exposed the falsity of their arguments, "put them to shame". Her action so vexed the tsar that he ordered half of her ancestral estate confiscated. Things might have gone worse for her had it not been for the intercession of the tsar's first wife, Maria. During the period of reprieve which followed, reports the author (again in the hagiographical tradition), Theodosia performed numerous "feats", giving away much of her property to the poor, redeeming the debts of persons condemned to *pravež*[4] and sheltering lepers in her home. To these deeds were added traditional feats of prayer and fasting, after Theodosia had placed herself under the spiritual guidance of Melania, an Old Believer nun. Such "feats" were petty formalities compared with what was to come, however.

It is in the second confrontation that the author's skill with dialogue becomes apparent. The occasion was a visit to Theodosia by her elderly uncle M. A. Rtiščev and his daughter Anna, both supporters of Nikon's reforms. In presenting their conversation the author avoids certain pitfalls into which a less talented writer might have fallen. For one thing, he refrains from cluttering his dialogue with long disquisitions on fine points of theology. Instead, he sidesteps detailed theological discussions with such statements as "she had great disputes with them", or "much else did she say concerning piety", or "with God's help she put them to shame". The author has drawn upon a variety of images and rhetorical devices, but he employs them with moderation, as though realizing that their excessive use would detract from the credibility and genuine pathos of his tale. Most important, however, is the fact that the author, despite his unflagging admiration for Theodosia and his contempt for Nikon's followers, seems to have presented the gist of what was said on both sides as accurately as he could. The utterances of Theodosia's interlocutors do not serve simply as cues or springboards for her declarations. They are almost always credible; often they are colorful and emotionally charged as well. Already in the second confrontation, after Theodosia's aged uncle has failed to persuade her that Avvakum has harmed and deceived her, the uncle's daughter Anna laments that the schismatics

[4] *Pravež* has been described by an eyewitness as follows: "The *praveush* [sic] or righter, is a place near the office: where such as have sentence passed against them, and refuse to pay that which is adjudged, are beaten with great cudgels on the shins and calves of their legges. Every forenoone, from eight to eleven, they are set upon the praveush and beate in this sort until the monie be payed." Giles Fletcher, *Of the Russe Commonwealth* (London, 1856), p. 67.

have consumed thee; they have swallowed up thy soul; like a little bird have they taken thee from us! Not only hast thou come to despise us, thy kin, but thou findest no joy in thy son. Thou hast but this one child, and thou lookest not upon him. And yet what a child he is! Who will not marvel at his beauty? It would be fitting for thee, when he sleepeth, to sit at his side and hold watch, and to place a candle of purest wax and keep a lighted lamp above the beauty of his form, and to glance at his face, and to rejoice that God hath presented thee with such a son! Often hath the Tsar himself, with his wife, greatly wondered at him and his beauty. But for thee he hath little worth and thou dost not heed our great sovereign. And if, because of thy stubbornness, the Tsar's burning wrath falls upon thee and thy house ... thou shalt have caused much grief and made thy son a beggar through thy mercilessness.[5]

To this Theodosia denies having been led from the path of true faith. She also claims to love her son, but

am I, for the love of my son Ivan, to harm my own soul, or for pity of my son, to renounce piety?...Nay, I wish not, in sparing my son, to destroy myself even though he be my flesh and blood, for I love Christ more than my son. Let this be known to you: if ye plan to bar my path to Christ by means of my son, ye shall never succeed! ... Even if ye put my son to fire, or give him to dogs to be torn to pieces in seeking to make me give up my faith, I shall not accept the new belief ... I shall remain in Christ's faith to the end, and shall prepare to taste death for His sake.[6]

Shortly after Theodosia's sister, Princess Urusova, had joined her in the ranks of the schismatics, Theodosia became a nun, taking the name Theodora. Meanwhile the tsar's first wife had died and he had remarried. Theodora's refusal to attend his wedding (she could not face the prospect of addressing him as "pious" [*blagovernyj*] at the ceremonies, so she plead that she could not come because her legs hurt) had infuriated the monarch. This gave rise to a third confrontation (more accurately, a series of two confrontations occurring a month apart) which took place when he sent the boyar Troekurov and Prince Urusov (Theodora's brother-in-law) with reprimands. The reprimand which he made through Prince Urusov was coupled with a threat: if she did not accept the new ways, great grief would befall her.[7] Her reply to the envoys was that she had done nothing wrong. Affecting amazement that the tsar should be angry "at my humility", she declared that it was not befitting for a sovereign to "persecute me, an humble woman". She would never, in any event, renounce "our Orthodox faith".

This response enraged the tsar, and the fury of the Nikon faction mounted as

5 Subbotin, *op. cit.*, 151-152.

6 *Ibid.*, 152-153.

7 The passage is interesting as an example of reported speech or indirect discourse, which is infrequently encountered in Old Russian literature: съ выговоромъ же, еже бы покорилася и пріяла бы новоизданныя законы ихъ.

Morozova continued to "expose their erroneous nonsense"[8] wherever she went, until they wished, "like savages, to devour her alive".

The author employs several devices, retardation in particular, to develop an atmosphere of anxiety and suspense preceding the fourth confrontation. The ominous possibility of a visit by an official search party is only suggested at first. A group of schismatic nuns had been living with Theodora and bringing her great comfort by their company; she was now urged by unnamed sympathizers to send the nuns away lest they be seized in her home. At first Theodora disregarded this advice and — in intimate vernacular — reassured her friends that there was no danger yet. But on the fourteenth of November she announced, without explanation, that her time had come. (Had she been informed, or does the author wish to imply that she, like Christ and some of His apostles, possessed the power to foresee coming events?) She bade farewell to her beloved coreligionists. On Shrovetide her sister, Princess Urusova, who had also been staying with her, went home. Prince Urusov secretly sympathized with Theodora and asked his wife to give "her friends" a message — a scriptural quotation. Princess Urusova rejoiced at his words but must have recognized the cheerless prophecy which they contained:

> for they will deliver you up to the councils and they will scourge you in their temples, and ye shall be brought before governors and kings for my sake.[9]

Prince Urusov spoke of impending woe for Lady Morozova, but seemed unaware of the extent to which his own wife was involved with the schismatics. One morning when she asked his permission to go visit Theodora, he replied that she should go to bid her farewell but should not tarry at her house, because he suspected that "they are going to send someone to her today". Against her husband's advice, Princess Urusova remained in her sister's home until night to await the "promised guests". In the "second hour of night" the women heard the main gates open. Morozova was suddenly filled with terror and knelt over a bench (in a posture suggesting Christ's moment of lonely anguish in the Garden of Gethsemane?). Her sister bade her take courage: "Mother-sister, be brave! Christ is with us — do not fear! Arise, let us make a beginning!" Each of the women then made seven bows, blessed each other "to bear witness to the truth", and prepared to face their tormentors. The stage was set for the fourth confrontation.

The confrontation is described with enough detail as to setting, lighting, location of the speakers and their posture and reactions so that we may visualize a series of inquisition tableaux. The archmandrite Joachim, who had come with

[8] блядство заблужденное — an example of paranomasia?

[9] Matthew 10:17-18.

henchmen to question Morozova, "brazenly entered the blessed one's bedroom". He began to interrogate her harshly. When he ordered her to show him how she crossed herself, she did so in the schismatic manner. But, to emphasize her contempt for the tsar in whose name Joachim was questioning her, she refused to rise from her bed and she would not disclose the whereabouts of her co-worshippers. Meanwhile Joachim's aide, the councilor Ilarion, accidentally came upon Princess Urusova, who was lying in an adjoining, unlit storeroom. Upon learning her identity he "jumped back, horrorstruck, as though burned by fire", and balked at the archmandrite's orders to make her cross herself. "We were only sent", argued Ilarion, "to see Lady Theodosia." The irked archmandrite insisted that the princess be "tested" also. When Princess Urusova revealed herself as Morozova's coreligionist, the archmandrite interrupted proceedings to go and make a personal report to the tsar. He then returned to interrogate Lady Morozova's maidservants. "A black sexton named Joseph, standing outside the door", directed the archmandrite to question two servant girls in particular, identifying them by name. Both admitted sharing their mistress' beliefs. Lady Morozova and Princess Urusova were arrested and taken from the house. As Morozova was being carried out (she refused to leave voluntarily) her young son came out and "bade her farewell from behind". She did not notice him, however, and he went back. Theodora and her sister were both put in chains and placed in a cellar under guard.

The next confrontation followed two days later when the principal interrogator was Metropolitan Paul (Krutickij). Starting in a restrained tone, the interrogation turned into an impassioned debate, conducted in extremely lively dialogue. Under the metropolitan's prodding Theodora declared that, though the tsar and his family were worshipping according to Nikon's "corrupt service books", *she* refused to do so. When the Metropolitan asked, "What then dost thou think of us — that we are all heretics?" she gave an answer worthy of Avvakum himself: "Since God's foe, Nikon, has thrown up his heresies like vomit, and ye are now licking up his profanations, it is clear that ye are just like him." Hearing these words, Metropolitan Paul and the others "began to shout loudly". Archmandrite Joachim cried out that it was wrong to address her as "mother" and "venerable one" — or even as her father's daughter; in truth she deserved only to be called the daughter of the devil. The debate lasted "from the second hour of night until the twelfth hour".

The sisters were returned in chains to their cellar but the next day Theodora was taken away on a sledge (Surikov's painting presumably depicts this scene), to be followed by Princess Urusova. At a moment when she thought the tsar might see her from his window, Theodora sat up in her sledge and defiantly clanked her chains, lifting her hand high, with fingers raised in the schismatic manner. The sisters' new prison was the Alekseev convent, from which they were periodically "dragged" to attend church services. Since Theodora resisted going to these services by lying down and refusing to move, it was necessary to transport her back and forth on a stretcher "like a dead body". When she passed people she knew on the street, however, she would stop her stretcher-bearers by a ruse ("Alas, I have become exceedingly weary; stop for a moment!") then she would vociferously berate them for compelling her to attend services which she shunned. Thus she made her attitude clear to all who saw her in the streets. She was soon transferred to another convent and placed under heavy guard. Here she was joined by Princess Urusova and another prisoner, a co-worshipper named Maria, the third of the "martyrs" whose names appear in the title of the work. Theodora now received visits from another Metropolitan (Ilarion of Rjazan'), but we are told only that she "argued so cleverly that he was greatly put to shame and often had to leave without being able to answer".

Theodora's son Ivan, left at home, pined for his mother and fell ill. The tsar sent his own doctors to look after the boy; they, says the author cryptically, "only treated him in such a way that he was delivered to the grave a few days later". A priest, sent to tell Theodora of his death, quoted from the psalms to show that her son had died as a result of her deserting, Judas-like, the true faith; this was God's punishment. Theodora took the word of Ivan's death very hard. Falling to the ground, "she cried out in a tender voice, weeping and sobbing: 'Woe unto me, my child! The apostates have destroyed thee!' And for a long time she did not rise from the ground, chanting funeral songs for her son."[10]

Following the son's death, Tsar Aleksej Mixajlovič confiscated all of Morozova's property, though subsequently he permitted her and her sister to have maidservants in prison. The third "martyr", Maria, was less fortunate and was subjected to "vile acts" by the "shameless soldiers" who were guarding her.

Many noblemen's wives and simple folk would gather to watch the ladies being forcibly transported to church services. Worried by the large crowds, the abbess reported the matter to the new patriarch, Pitirim. Pitirim had somehow known virtually nothing about the Morozova case. After listening with great interest to the abbess' account, the new head of the Russian church went to the tsar with a recommendation that Morozova be set free and some of her property restored to her. It would appear more proper, opined the patriarch, to handle the case with leniency. This foolishness was a "woman's matter"; who expected women to understand doctrinal issues, after all? The tsar replied that he would have been happy with such a solution, but that the patriarch did not know "that woman's ferocity". No one had caused the tsar as much woe as

[10] For some eighteenth-century versions of the life which have greatly expanded this and other scenes, see Mazunin, "Ob odnoj pererabotke", pp. 430-431.

she. How she had railed at him, and how she continued to abuse him! If the patriarch did not believe him, he could go see for himself. Let the patriarch summon her and question her; "then shalt thou learn her steadfastness, and when thou questionest her, then shalt thou taste her spiciness". Whatever the patriarch decided to do would meet the tsar's approval.

Morozova was summoned "at the second hour of night" to the ecumenical chamber of the Miracle Monastery, where she found herself in the presence of Patriarch Pitirim, Metropolitan Paul and other high-ranking dignitaries, both ecclesiastical and civil. The confrontation is one of the most dramatic in the whole *vita*. Pitirim regarded Theodora as a misguided child. Making no progress in his preliminary arguments with her, he decided to take direct action: "because I am much concerned with thee, I myself, in my old age, shall take thy confession ... and shall administer to thee the sacrament". Morozova had refused to stand while conversing with the patriarch, but when the elderly pontiff approached her in order to administer the sacrament and anoint her with holy oil, she rose to her feet and prepared herself "like a fighter". We shall not describe all the moves and countermoves in this fine scene. Enough to say that she repulsed her adversaries, forcibly pushing away the patriarch's hand at the culminating point and crying aloud: "Destroy me not with thy apostate oil!" Taking her chains in hand, she asked: "For what have I worn these chains all year ...? Thou wouldst in a single hour bring to naught all my suffering and toil! I shall never have need of thy holy things!" The patriarch retreated, "roaring like a bear" and shouting to those present to "throw her down, drag her mercilessly, like a dog! Chain her by the neck! Drag her away from here! She is the daughter of the devil ...! She shall live no longer! Take her to be burned!"

Princess Urusova and Maria were similarly brought before the patriarch that night. They likewise acquitted themselves with courage — and with intriguing varieties of behavior calculated to shame their interrogators. Princess Urusova, for example, removed her headdress, to the panic of all present. The patriarch, "unable to bear the dishonor", reported all to the tsar, but complained most passionately about Morozova. "Did I not tell thee of that woman's ferocity?" replied the monarch. "Thou hast seen her deeds but once, while I have suffered from her for years."

The tsar then ordered the women tortured into submission. His order was carried out the following night in the crowded courtyard of the communications building. There are few scenes in Old Russian literature which can approach the graphically-detailed horror of the scene which followed.

Before discussing the torture scene we should devote a few lines to the author's methods of developing the conflict and tension which preceded it. His attention to details of time ("in the second hour of night") and place ("she was transported past the Miracle Monastery near the Tsar's Passage") give events a background of almost documentary reality — an effect which the author may not have sought consciously, desiring merely to record key facts. The effect of credibility is heightened by the narrator's ability to keep his personal viewpoint in check. His own sympathies are totally on the side of his heroine. He knew her well (visiting her at least once during her incarceration) and speaks of her in a tone of unflagging admiration — a tone sustained by epithets (*blažennaja, preblažennaja, dobljaja, revnitel'nica, prepodobnaja, premudraja, svjataja, velikaja, mučenica*), word-compounds (*prisnopamjatnaja, strastoterpica, svjatejšaja strastonosica*) and metaphors (*adamánt, neporočnaja agnica, bezcennyj biser*). Such a viewpoint is not surprising in old Russian literature. What is surprising is the author's treatment Morozova's opponents. Since he was the apologist for a persecuted "saint" in a period of religious turmoil, and since he considered the Nikonites to be in league with the devil, the author might have been expected to portray Theodora's adversaries in a one-sided, distorted manner. Yet this is really not the case. Aside from some denunciatory epithets cast at his heroine's antagonists, the author seems to have tried to present them accurately. Perhaps he was convinced that the events would speak for themselves. In any event few of Morozova's foes bear any resemblance to agents of the devil or to heartless malefactors — at least up to the torture scene. The author endows them with very human emotions and reactions to their "misguided" sister. Early in a confrontation they may exhibit typically Christian motivations. The patriarch, for example, seems to have felt pity for Theodora at first, and to have wished to help her.

But the author's development and intensification of the conflict are much more than the result of factual reporting or a credible view of the struggle. The account of the conflict in its early stages, we recall, took the form of a statement of the tsar's displeasure and his partial confiscation of Morozova's property after she had refused to heed his counsel and had "shamed" his envoys. These events are recorded laconically, without dialogue. Next came private remonstrances and warnings from relatives — the aged uncle and his daughter Anna. Both expressed themselves in a shocked, grieved dialogue. As the conflict progresses through a series of confrontations, the narrator portrays the frustration and rage of Morozova's interlocutors more and more vividly: Councilor Ilarion is vexed but hesitant, while the archmandrite Joachim is angry and "arrogant"; the metropolitan and his associates are moved to shouting and to abusive epithets ("devil's daughter!"); finally the screaming patriarch orders her dragged mercilessly "like a dog", and demands the death penalty for her.

The torture scene describes how the three women were humiliated by being stripped to the waist, tormented by being "brought to the fire" and hoisted on the *trjaska* — a wooden rack-like instrument which tortured by shaking the victim violently. Morozova, who defiantly replied to her tormentors' verbal abuse and compared them to Christ's tormentors, was kept on the *trjaska* longer than the others. All three women were then cast into the snow and kept there three hours, while their adversaries perpetrated additional indignities on them. The high point of dramatic tension is reached, in our opinion, at the end of the torture scene, after Maria had been mercilessly flogged front and back, and Morozova finally screamed: "Is this Christianity — to torture a human being thus?" The women were taken away "in the ninth hour of night".

On the following day the tsar summoned a council to determine the fate of Lady Morozova and her fellow schismatics. The patriarch urged that Morozova be burned alive; a wooden framework (*strub*) was even readied for the purpose in a marsh. Some of the boyars opposed such a course of action. Melania saw the wooden framework in the marsh and informed Theodora and her sister. Both welcomed death, looking forward to meeting their Lord in heaven. But God had other plans. He willed that "the lady martyrs suffer yet a long time" on earth. The struggle goes on, but the tactics change, and with them the mood of the story.

The main protagonists are, of course, Lady Morozova and the tsar, and it seems to us that the author, in depicting their struggle, has succeeded in creating two characters of a depth and "roundness" rarely encountered in Old Russian literature. His success is the more remarkable because of the limited varity of methods which he employs. There is no "lyrical monologue" (of the type found in the *Passion and Encomium of SS Boris and Gleb the Martyrs*, for example) and the time had not yet come when Russian authors made extensive use of physical description as a method of characterization; if we try to visualize Morozova, for example, we are apt to see her as Surikov painted her in the late 19th century; certainly the *Life* does not tell us what she looked like. Yet early in the story it becomes clear that Lady Morozova is a far cry from most of her stylized, kenotic predecessors and even from Juliana Lazarevskaja, with whom she is sometimes compared.[11] The various sides of Morozova's personality emerge from what she herself says, from what others say to or about her, and from what the author says in straight narrative. Theodora could be earthy and abusive when holding forth against adversaries ("Nikon has thrown up his heresies like vomit, and ye are now licking up his profanations ...") but extremely tender and affectionate with friends and coreligionists. This affection is mirrored in the intimate vernacular which she uses with those she loves ("*Ni, golubicy moi, ne bojtesja*!"). Her courage and fortitude astound, but she had her moments of hesitation and weakness as well (for example, while awaiting

[11] Stender-Petersen, *Anthology of Old Russian Literature*, p. xvii.

the nocturnal *prisylka* of which Prince Urusov had warned). For all her fanatic bravado concerning her son Ivan, she was overcome by motherly grief when he died. There are other suggestions of the eternal feminine in Lady Morozova. She exerted a strong attraction on men as well as her female coreligionists. One variant quotes words of her brother-in-law, the great boyar Morozov, which sound like a love song.[12] Her hostility and repugnance for the Nikonite hierarchy is balanced by devotion to her spiritual father Avvakum and love for her "bridegroom" Christ, whom she insists on meeting in unsoiled clothes, if possible. We may, however, discount the testimony of some of her servants, quoted in another source, that she had been Avvakum's mistress.[13] Still other attributes of her character remain a puzzle. Why, for example, did she answer with a white lie ("my legs pain me so greatly that I can neither stand nor walk") when the tsar personally requested her presence at his wedding? Was she not fully prepared, while her son yet lived, to face the consequences which a more forthright explanation would have brought on her? Or did she wish to irritate the tsar with a petty excuse? According to the author she realized that he would not believe her and would be enormously displeased, as indeed he was. Her subsequent protestations of humility and innocence ("I know not of any evil that I have done to the tsar and am amazed at the tsar's anger at my humility ... It is befitting that the tsar not persecute me, a humble woman") may have had their origins in anxiety; or were they calculated to shame the monarch? Such maneuvers (or hints of frailty and compromise) were in any event absent from Theodora's later behavior when her son had died and her back was to the wall.

The author presents the tsar, Aleksej Mixajlovič, as a vacillating personality, and we should perhaps devote a paragraph to the characterization of the monarch before proceeding with our discussion of the narrative. Aleksej Mixajlovič could be gentle and kind: he was fond of Theodora's son Ivan and was reluctant to take harsh measures against the mother as long as the son was alive. (One of the less convincing parts of the *Life*, in our opinion, is the passage which suggests that the tsar's doctors were partly responsible for Ivan's death.) His behavior seemed indecisive at times, and his frequent consultation with his boyars on what to do with Morozova again suggests a reluctance to deal brutally with her on his own initiative; she was, after all, a noblewoman. But he could be vindictive; he appeared more upset by her refusal to attend his wedding than by her schismatic religious views, and his struggle with her had definitely personal undercurrents.

[12] Subbotin, *op. cit.*, p. 146, footnote 1.

[13] For a discussion of this and other aspects of his relations to Lady Morozova, see Avvakum's account in "O trex ispovednicax. Slovo plačevnoe", in *Žitie protopopa Avvakuma im samim napisannoe i drugie ego sočinenija*, ed. N. K. Gudzij (Moscow, 1959), pp. 295-302. The same volume contains Avvakum's letters to Lady Morozova: pp. 207-217.

As the story moves on, the tsar gives evidence of feeling uncomfortable — possibly remorseful — about the persecution of Lady Morozova and adopts a conciliatory tone. Three days after the torture he sent a musketeer officer to her with a message couched in terms of pleading and admiration. Morozova was addressed as "venerable mother"; she was a second Catherine the Martyr; he wished to restore her to her original position of honor, if she would but make it possible for him to do so. If she would go through the motions of recanting, he would send his royal winter carriage, drawn by Asiatic horses, for her, and many boyars would come to "carry her in triumph on their heads". Lady Morozova rejected this truce offer. She had ridden before in winter carriages; she herself had owned Asiatic horses — and Tatar horses, too — but now such transitory possessions interested her little. Her only desire was to receive "Christ's gift" of fiery death and reunion with her Lord.

An event soon occurred to which the author ascribes a supernatural cause: "God's justice soon befell Patriarch Pitirim and he ended his life with a dreadful death." The tsar ordered Lady Morozova moved to the Novodevičij convent, where she continued to be kept under guard and dragged to church services. But presently, vexed that noblemen's wives were flocking there in such numbers as to block the convent grounds with their carriages, the tsar ordered Theodora brought back to the city. His older sister chided him for this: "Why, dear brother, doest thou such unseemly things, ordering that widow moved about from place to place? It is not good, dear brother!" And she bade him remember Boris and Gleb. "Fine, dear sister, fine!" shouted the tsar in great wrath (and in vernacular Russian), "If thou art patronizing her, I have a place ready for her now!" Morozova was sent to Borovsk, about seventy-five miles to the southwest of Moscow.

In Borovsk Theodora and her fellow-sufferers were ultimately deprived of almost all their belongings, (including Morozova's prayer books and icons, to her immense distress) and were placed in a deep dungeon without food or water. Anyone trying to visit or communicate with them risked the death penalty. The author graphically describes their sufferings in the "unlit darkness": their hunger and thirst, the suffocating atmosphere, the lice. For a while the women somehow subsisted on food and water secretly lowered to them by "Christ-loving" soldier guards. Princess Urusova survived this existence for ten weeks, then died. As her body was lifted out by rope, Morozova wept and bade farewell to her "most beloved flower". A traditional *vita* element appears when the author assures us that Princess Urusova's body lay "where it had been thrown" for about five days while prison authorities awaited word from Moscow on disposition of the corpse, and "not only did it not darken, but each day did it grow brighter and whiter".

A final confrontation with the enemy occurred when a Nikonite monk came to make a last attempt to change Lady Morozova's mind. But she had worn her chains ("recalling the chains of Paul") for four years; her sister had already gone to rejoin her Lord, and Theodora was herself ready to meet her "good and beautiful Lord". According to this account, the monk left in tears, praising(!) her and the cause for which she was ready to die.

Morozova's last conversations — it would hardly be appropriate to call them "confrontations" — were with a frightened soldier whose duty was to guard her:

After this the most-blessed Theodora became extremely ill and called one of the soldiers and said unto him:

"Slave of Christ! Dost thou have a father or mother living, or have they died? For if they are alive, we shall pray for them and for thee; if they are dead we shall pray for their souls. Have mercy, slave of Christ! I have grown weak from hunger and I long to eat. Have mercy, give me a little cake!"

But he said:

"No, my lady, I am afraid."

And the martyr said:

"At least a little bread."

And he said:

"I have none."

And again the martyr [spoke]:

"At least a few biscuits."

And he said:

"I have none."

And Theodora said:

"If thou dost not dare, then bring me at least a little apple or a tiny cucumber."

And he said:

"I do not dare."

And the blessed one said:

"All right, child! Blessed be our God who has willed it thus! And if it is really to be as thou hast said, then I beseech thee to do a last act of love: place this humble body of mine, covered with a mat, near my beloved sister and fellow-sufferer, inseparably [from her]."

Then, when she had reached the last stage of weakness, she again called the soldier and said unto him:

"Slave of Christ! Didst thou have a mother? I know that thou wast born of woman. For this reason I beg this of thee, guarded by the fear of God: for I am a woman, and, being oppressed by great hardship, I need to have my underclothing washed. As thou thyself canst see, it is impossible for me to go and serve myself, for I am chained here and have no maidservants to attend me. So go thou to the river and wash out this undergarment of mine. For the Lord is about to take me from this life, and it is not fitting for me to lie down wearing unclean clothes in my mother earth."

And having said this, she gave him her rags. And, hiding them under his garment, he went and washed them in the river. And as he washed the little fabric, his own face was bathed in tears ..."

Lady Morozova "peacefully went to sleep" (in the *žitie* manner) "on the second of November in the first hour of night". She was taken to be buried, as she had requested, near Princess Urusova, "and was wrapped in a mat instead of costly silks". On the night of her death, the author tells us, her spiritual mother Melanija was in the "wilderness" and had a vision in her sleep. Whether this was a vision in the supernatural tradition of hagiography or a dream of the more modern literary type, it must have brought deep satisfaction to Old Believer readers, for it described Theodora dressed as a nun in a "wondrous cowl, and Theodora was bright of face and rejoicing, and she was beautiful in the cowl in her happiness, and looked all about herself, and was drawing her hands over her clothes and wondering at the beauty of her vestments and constantly kissing the Savior's icon which was near her, and also kissing the crosses on her nun's garment."

Melanija's dream-vision provides a sort of heavenly denouement for the tale. Instead of proceeding with a lengthy and rhetorical encomium of Lady Morozova and her sisters-in-suffering (something we could anticipate in a *vita*), it concludes with a few lines devoted to the third martyr, Maria:

> Maria lived only a short time after Theodora, just one month: on the second day of December she presented herself to her Lord. And the third [woman] ascended to join the other two in everlasting rejoicing in our Lord Jesus Christ to whom belong all glory, honor, worship and majesty, with His supreme Father and the Holy Ghost, now and always, for evermore. Amen.

It could be argued that this Old Russian *vita* holds our interest because of its theme — the theme of the individual who suffers persecution by the authorities for holding officially-proscribed views. There is nothing quaint or outmoded about this theme; it is as gripping today as it has ever been. In the last analysis, however, we would be more inclined to attribute the moving qualities of Lady Morozova's tale to the author's narrative skill, certain features of which we have discussed in this paper: his avoidance, on the one hand, of theological disquisitions, long prayers, and other lyrical digressions once the story gets under way; on the other hand, his selection of detail, use of retardation and other devices to develop suspense and conflict, his methods of characterization and the effective dialogue (especially in the story's numerous confrontations). It is a good story, well told and deserving of more attention as a literary work than it has received so far.

LESKOV'S TRIP ABROAD IN 1875
(*Four Unpublished Letters to I. S. Gagarin*)

WILLIAM B. EDGERTON

Indiana University

The year 1875 was a turning point in the life of the Russian writer Nikolaj Leskov. During the previous winter, after three highly profitable years of publishing in *Russkij vestnik*, Leskov had broken off relations with the journal because of conflicts with the editorial staff over his latest work, *Zaxudalyj rod.* Having earlier been proscribed by the liberal and radical press because of his newspaper articles on the Petersburg fires of 1862 and especially because of his sensational novel *Nekuda* in 1864, Leskov himself, like the heroes of the novel, now almost literally had "nowhere to go". During the winter of 1874-75 he tried in vain through I. S. Aksakov and other acquaintances to get a commercial position in the business enterprises of the wealthy Moscow Old Believer Vasilij Aleksandrovič Kokorev, hoping in this way to free himself from the "overwhelming oppression of the journals"[1] so that he could write as he pleased. By mid-April 1875 he had decided to go abroad again. "For a while at least", he wrote on April 23 to I. S. Aksakov, "I want to avoid seeing all that has deprived me of strength and activity. I plan to join some group of French pilgrims and go along with them to Lourdes. Perhaps that religious awakening of people whom I know for their irreligion will arouse my interest and keep me from thinking thoughts that are so painful and fruitless. Beyond that, I don't even know just why I am going; I only feel an irresistible urge to get away, and for as long a time as possible."[2]

On his way to Western Europe Leskov stopped in Moscow and spent a day with I. S. Aksakov at his country house in the village of Kuncevo. Aksakov asked him to call on the Russian Jesuit priests Ivan Sergeevič Gagarin and Ivan Mixajlovič Martynov in Paris, with whom the Moscow Slavophiles had engaged in some polemics in the 1860's on the question of Jesuits in Russia.[3]

[1] Letter to I. S. Aksakov, November 16, 1874, in Н. С. Лесков, *Собрание сочинений в одиннадцати томах* (Москва, 1956-58), X, 362; hereafter cited as *Собрание сочинений.*

[2] *Собрание сочинений*, X, 397.

[3] *Собрание сочинений*, X, 418. In his article "Иезуит Гагарин в деле Пушкина", *Исторический вестник*, VII (август 1886), 269-273, Leskov describes his visit to Kuncevo, during which they talked about the old rumor that Gagarin was the author of the anonymous

From Moscow Leskov proceeded to France by way of Kiev and Vienna, arriving in Paris some time before June 11 (May 30, Old Style). His original plans to join French Catholics in a pilgrimage on foot to Lourdes were abandoned,[4] and he remained in Paris until his departure on July 15/3 for Marienbad.

During these two months in Paris it is clear that Leskov's most important contacts were with Fathers Gagarin and Martynov.[5] Information about the nature of their associations can be found in four unpublished letters and a copy of a fifth from Gagarin to Leskov which are preserved in Moscow in the Central State Archive of Literature and Art and in four letters from Leskov to Gagarin now preserved in Paris in the Bibliothèque Slave, which was founded by Gagarin himself. I am deeply indebted to the eminent French Slavist Professor André Mazon for his help in finding these letters from Leskov, and to Father M. J. Rouët de Journel, S. J., the director of the Bibliothèque Slave, for his kindness in making them available to me for publication.

Leskov's first letter was written in reply to a note from Gagarin dated June 11 [May 30], 1875, in which Gagarin offered to take him on a tour of the Jesuit

---

lampoon in 1836 naming Puškin "coadjuteur du Grand-Maître de l'Ordre des Cocus". Aksakov said he believed the rumor was false. Leskov reached the same conclusion in Paris on the basis of his acquaintance with Gagarin. Final vindication for Gagarin came only in 1927, when Soviet handwriting experts proved that the author of the lampoon was Petr Dolgorukov. Cf. П. Е. Щеголев, *Дуэль и смерть Пушкина*, изд. 3-е просмотренное и дополненное (М.-Л., 1928).

4 Leskov gave a variety of reasons for changing his mind about this pilgrimage. In a letter to his children on July 12 [June 30] he said the Garonne river had flooded the region just as he was getting ready to go. But six days earlier he had written to A. P. Miljukov: "I shall not go to Lourdes, both because it costs a lot and because I see and understand very clearly what those pilgrimages are." On August 10/July 29 he wrote from Marienbad to I. S. Aksakov that he had planned to go to Lourdes "but Lourdes at that time was flooded by the Garonne and I was flooded by a no less overwhelming melancholy, which was beyond description" (*Собрание сочинений*, X, 404, 406, and 415). In the *Собрание сочинений* Leskov's letter to his children, mentioned above, is wrongly dated "June 12 (24) 1875". In his *Жизнь Николая Лескова* (M., 1954), p. 314, Leskov's son Andrej Nikolaevič correctly dates the letter July 12 but does not indicate whether this date is Old Style or New Style. The date July 12/June 30 is proved correct by the street address on the letter as published by Andrej Leskov, 5, rue Chateaubriand (on June 24/12 he was still living at 9, rue Monsieur le Prince); by the reference in this letter to his having been abroad for two months; and by his statement that he planned to leave for Marienbad "the day after tomorrow", in other words, on July 15/3. That he really did leave on that date is proved by his letter of July 22/10 to his son Andrej saying that he has been in Marienbad "one week today" (*Жизнь Николая Лескова*, p. 316).

5 Prince Ivan Sergeevič Gagarin (1814-82) was acquainted with a number of leading Russian literary figures, including Puškin, Lermontov, and Herzen, and was greatly influenced by Čaadaev. In 1843 he emigrated to France, was converted to Catholicism, and soon became a Jesuit. Ivan Mixajlovič Martynov (1821-94) emigrated to France in the 1860's and shortly thereafter became a Catholic and a Jesuit. Both men published many works dealing with religious matters, and Martynov was also active as an archeologist.

schools in Paris.[6] Leskov's answer shows that this exchange of letters does not mark their first contact with each other. He refers to a conversation with Martynov and to books that Gagarin has given him, and he expresses appreciation for the kindness that Gagarin and Martynov have shown him.

No. 1. June 13/1 [?], 1875. Paris.

Paris. 9, rue Mr le Prince, 11 Juni 1875 г.[7]

Мой отец!

Не умею благодарить Вас за Ваше незаслуженное мною внимание: видеть иезуитские школы для меня, как для писателя и как для члена ученого комитета министерства, величайшее удовольствие, какое только мог бы мне кто ни будь доставить. Я буду с нетерпением ждать Вас в 1 час, завтра (в понедельник.)

В ту минуту, когда я получил Ваш[у]*[8] записочку я сам хотел писать Вам (д[а]* по правде сказать уже и написал). Ув[а]жаемый* отец Мартынов сказал мне [что-]то* весьма меня заинтересовавшее логичностью своего построения и я хотел бы с ним поразуметься. Вы оба очень добр[ы]* и прозорливы: меня отнюдь не занимают веселости Парижа и от них я спешу бежать; но вопросы веры, Церкви и школы меня интересуют неустанно и Вы будете для меня незабвенны если дадите мне испить от сего источника.

Одну из подаренных Вами мне книг, — именно "*Le clergé russe*" я пробежал и о ней-то и хотел писать Вам. В ней очень много справедливого, — особенно вывод на стр. 288 (начиная от слов "L'Eglise" до слов "corps sans âme";)[9] но ка́к Вы говоря о монастырях могли забыть наших удивительных подвижников, им же несть числа!... Зачем Вы не засвидетельствовали о высоких душах, доселе кроющихся в горах, вертепах и пропастях земных!... Без них нет картины русского иночества, а есть только очерк архиерейских домов и ставропигий. Или Вы не знали Серафима Саровского,[10] Парфения Голосеевского,[11] Макария Оптинского[12] и дивных

[6] Центральный государственный архив литературы и искусства (ЦГАЛИ), ф. 275, оп. 1, ед. хр. 218.

[7] This date is bound to be a mistake, because Leskov refers to "tomorrow" as Monday. In 1875 June 11 fell on Friday. Evidently Leskov wrote this letter on Sunday, June 13/1.

[8] This letter was evidently damaged on being opened. A strip of margin is missing from all four pages. The asterisks indicate emendations along this margin.

[9] Le Père J. Gagarin, de la Compagnie de Jésus, *Le Clergé russe*. Nouvelle édition (Bruxelles: Librairie Pontificale de H. Goemaere, 1871). The passage to which Leskov referred is the following: "L'Eglise n'a pas de vie qui lui soit propre: en toutes choses elle reçoit l'impulsion du dehors. Le clergé porte des mitres et des chapes; c'est la seule chose qui le distingue des autres fonctionnaires de l'Etat. L'Eglise russe n'est pas asservie, elle est absorbée par l'Etat; c'est un instrument inerte, un corps sans âme."

[10] Serafim Sarovskij, secular name Proxor Sidorov Mošnin (1759-1833) — elder of the Sarova Hermitage, in Tambov Province.

[11] Parfenij Goloseevskij, secular name Petr Krasnopevcev (1790-1855) — a monk at the Kievo-Pečersk Monastery, who was the father confessor and friend of the Kiev Metropolitan Filaret.

[12] Makarij Optinskij — probably the fourteenth-century founder, according to tradition, of the Optina-Vvedenskaja-Makarieva Hermitage, the former leader of a band of robbers,

отшельников Соловков и Валаама? Признаюсь, этот пробел меня удивляет.

Прочел я и Вашу русскую книжечку "*Сокровище христианина*"[13] и она меня тоже заинтересовала, как конкрет Р[имско] католических положений с которыми я не имею желанного, полного знакомства; но меня она поражает, во 1-х) тоном: Вы как бы коментируете Творца, на что человек едва ли имеет полномочие, а во 2) Вы так ригористически говорите о единственном грехе, что я не знаю: как Вы помирите эти страхи с 13-15 ст[ихами] 3-й гл[авы] 1-го посл[ания] Коринфянам?[14] Судите меня как хотите, но Бог в *Вашем* представлении страшен и ужасен, — это не евангельский Бог милосердия, а старый Библейский Бог, о котором и библейские-то писатели, или гугнявили как Моисей, или безмолвствовали, как немой Захария.[15] Считая себя мерзким и недостойным христианином, я не могу принять *такого* мстительного Бога. Про[шу]* Вас: Рассмотрите это место (1. Кор[инфянам] 3.13[)]* — куда Вы его денете? Не указывайт[е]* мне на И[оанна] Злат[оуста][16] — я знаю его изъяснения и не считаю их удовлетворительными. Н[е]* покажете ли мне что ни будь иное, а [до]* тех пор я буду в вере в Бога *м[и]лости* бесконечной*. Скажите мне: зачем это запугиванье: *кого оно испугало и где его добрые плоды*? — их нет и не будет. Русская вера идет иначе: я знал старца, который молился: "Господи! пошли меня в ад, — не дай мне отдыха, а дай его слабым и Тебя не знающим". Этот *дух* по моему ближе к духу евангельскому. И зачем эта коментировка? Право мне кажется, что это ужасное своеволие в деле судеб Господних.

"Простить Он может, хоть осудит."[17]

Зачем же Вы отнимаете надежду на манифест после решения? Или Вы противуречите молитве за усопших и всей идее чистилища.

Однако простите меня за неуместную может быть откровенность и многоглаголание.

---

who was named Opta; or else the nineteeth-century Father Superior of the hermitage (secular name Mixail Ivanov), who died in 1860.

13 *Сокровище христианина, или Краткое изложение главных истин веры и обязанностей христианина* (Париж в императорской типографии, 1855). Leskov's letter apparently serves as proof that Gagarin was the anonymous author of this rare little book.

14 I. Corinthians 3:13-15 — "Every man's work shall be made manifest: for the day shall declare it, because it shall be revealed by fire; and the fire shall try every man's work of what sort it is. If any man's work abide which he hath built thereupon, he shall receive a reward. If any man's work shall be burned, he shall suffer loss: but he himself shall be saved; yet so as by fire."

15 Zaxarij — Zacharias, the father of John the Baptist, who according to the New Testament account (Luke 1:5-79) was temporarily struck dumb as a sign that the angel Gabriel had spoken to him prophesying the birth of his son.

16 Ioann Zlatoust — Saint John Chrysostom (ca. 347-407), one of the greatest of the early Church Fathers.

17 "Простить Он может, хоть осудит" — a quotation from М. Ю. Лермонтов, "Демон", *Собрание сочинений в трех томах*, II (Москва-Ленинград, Издательство Академии наук СССР, 1959), 529.

Прошу Вас сказать мой поклон уважаемому отцу Мартынову и принять уверение в моем глубоком к Вам почтении.

Н. Лесков

On June 17 Gagarin sent Leskov a brief note saying that in response to Leskov's expression of interest he was enclosing a newspaper explaining the Catholic position on the worship of the sacred heart of Jesus.[18] Leskov's reply, written the same day, shows that he had somehow failed to receive the newspaper.

No. 2. June 17/5, 1875. Paris.

Мой отец!

Получил Вашу записочку и спешу благодарить Вас за внимание, которое уже ни как не смею приписывать ни чему иному, кроме большой доброты Вашей, мною ни мало не заслуженной.

Младая лань (и проч.) сосцы
млеком имея отягченны
нашла двух юных волченят
И стала исполнять долг матери священный.[19]

Приходится Вам быть "младою ланью", а мне волченком; но вот в чем дело: от газеты Вами мне посланной осталась "только одна зыбь поднебесная", а сама исчезла. Теперь что́ же в этом надо видеть: рок, коварство вражие, или просто, — что Вы ее не послали? Если не послали, то пожалуста пошлите.

Еще: черкните мне пожалуста Ваш водолечебный адрес. Что бы Вам было немножечко смешно я хочу распорядиться, что бы Вам *туда* была послана немедленно одна довольно смешная и забавная книга, называемая "*Смех и горе*".[20] Она пользовалась некоторым успехом, но до Вас по видимому не дошла.

Искренно Вам благодарный

Николай Лесков

17. июня [1]875 г.
Paris.

In his reply, dated "June 6/18, 1875", Gagarin explained that the newspaper he had sent was the issue of *Le Monde* for June 17, which was bound to be waiting

[18] A handwritten copy of this letter is in ЦГАЛИ, ф. 275, оп. 1, ед. хр. 218. The letter was published in A. C. Поляков, *О смерти Пушкина* (M., 1922), pp. 80-81.

[19] Quoted, not quite verbatim, from the opening lines of Krylov's fable "Лань и дервиш":

Младая Лань, своих лишась любезных чад,
Еще сосцы млеком имея отягченны,
Нашла в лесу двух малых волченят
И стала выполнять долг матери священный,
Своим питая их млеком.

[20] "Смех и горе", *Современная летопись*, № 1-3, 8-16 (1871), and republished in book form in the same year.

for Leskov at the concierge's. He also called Leskov's attention to a review of Tolstoj's *Anna Karenina* in the *Journal des Débats* for June 18. His closing lines indicate that their acquaintance pleased Gagarin quite as much as Leskov: "As for what you say about my kindness, I will simply tell you that it is always a pleasure for me to associate with Russians when they are not infected by insurmountable prejudices and nihilistic tendencies. And besides, from you one can learn much."[21]

On the following day Gagarin sent him another note saying that he would be at home all day and adding: "Tomorrow I plan to go to see you at one o'clock and show you another school."[22]

Leskov's answer to this, written the next day, is interesting not merely for its continuation of the religious dialogue with Gagarin but also for its references to the Russian painter Ge and to a "shaggy nihilist", who obviously was the prototype of Šeramur in the story by the same name published four years later.

No. 3. June 19/7, 1875. Paris.

19 июня [1]875. Paris.
9. rue M. le Prince.

Газета сегодня получена, — номер полон толкований о "sacré cœur" но это все не то чего я ждал, хотя впрочем я и не надеялся встретить ни чего иного. Мы все во всем Иисусе, — вся жизнь христианская может быть только в подражании Ему: "я в Вас и Вы во мне", — понятно что таким образом и святое сердце Его во всем участвует, — страдает и радуется, но к чему его особить, — я все таки не вижу и не могу этого понять. Потом: мне не нравится вообще увеличение числа праздников, которые приняли в христианском мире самый нехристианский характер, и это так повсюду, как на востоке, так и на западе. Праздники ныне синонимы *праздности*, имеющей оправдание в истории церковных событий, а не дни богомыслия, как это должно бы быть; а праздники такого сорта не могут быть богоугодны ("Празднований ваших и новомесячий не терплю, — *их ненавидит душа моя*" читаем у пророка.)[23] По моему учреждение празднеств не полезно, а вредно, и лютеране очень умно делают, что не творят их, как мы и римские католики. Надо, что бы все дни были божии; а у нас и у Вас, шесть дней работа и такое пьянство с обыденным развратом, а в седьмый день, благодаря праздности, пьянство ожесточеннее и разврат сугубый. В лютеранских странах *это* положительно лучше.

За обещание навестить меня и сводить в школу не умею Вас достойно благодарить, но вот в чем дело: здесь есть некий шаршовый нигилист,[24]

[21] ЦГАЛИ, ф. 275, оп. 1, ед. хр. 218.

[22] *Ibid.* This letter is dated only "Saturday". Its exact date is established by Leskov's letter of June 19 [7], in which he thanks Gagarin for the promise to take him to a school but explains that he does not know whether he will be back home from church the next day by one o'clock.

[23] Isaiah 1:14.

[24] The identity of this "shaggy nihilist" is unknown. He is unquestionably the prototype of

который умирает с голоду. Я его много накормить не могу, а хочу ему завтра после обедни выпросить кое что у русских, — по этому я должен быть в церкви и не могу в точности определить: когда вернусь домой. Позвольте мне просить Вас подождать меня дома, а я прямо из церкви, поговорив с кем удастся, приеду к Вам. Это будет гораздо вернее. Сегодня я провел 6 час[ов] на выставке картин и вернулся домой совсем без ног. Священная живопись во Франции положительно не имеет характера, хотя за то нет мерзких картин в роде "Тайной вечери" и "Воскресения" нашего г[осподина] Ге, который вместо воскресения изобразил трех римских воинов, считающих деньги.[25] Иконописи с ее вдохновляющим символизмом у Вас вовсе нет.

Искренно преданный Вам

Н. Лесков

On the following Tuesday, upon receiving two brochures from Gagarin, Leskov wrote the last of the four letters that are preserved in the Bibliothèque Slave:

No. 4. June 22/10, 1875. Paris.

22 июня [1]875 г. Paris.
9, rue M. le Prince.

Благодарю Вас, мой отец, за присланные Вами мне две брошуры, из которых одна мне была совершенно неизвестна, — это "исповедание веры для грекороссиян". Я не знал что такая брошура существует. Что же касается "любопытных свидетельств о непорочном зачатии св. Девы",[26] то они действительно очень курьезны и недвусмысленны. Особенно интересен мне Иоаникий Голятовский, с "Ключем" которого я не мало повозился.[27] Этот "Ключ" по моему не ключ, а так сказать вернее

Leskov's Šeramur in the sketch he first published in *Новое время*, Nos. 1352, 1359, 1361, 1362, and 1366 (December 2, 9, 11, 12, and 16, 1879), under the title "Дети Каина. Типические разновидности. Очерк первый. Шерамур. Эпизодические отрывки фатальной истории". This work has since been republished as "Шерамур". See *Собрание сочинений*, VI, 244-301.

25 Leskov's criticism here of the Russian painter Nikolaj Ge (1831-94) stands in sharp contrast to his opinion of Ge in later years, when their sympathy with Leo Tolstoj's religious views drew them close to each other. In a letter to his son on February 15, 1890, Leskov wrote of Ge's painting "Christ Before Pilate": "This is the *first* Christ I understand. It could only have been painted by Ge, the friend of Tolstoj" (Андрей Лесков, *Жизнь Николая Лескова*, p. 463).

26 *Исповедание веры для грекороссиян* ... Profession de foi présentée par le pape Grégoire XIII aux grecs qui veulent entrer dans la communion de la Sainte Eglise catholique, apostolique et romaine. Traduit sur le texte latin et imprimé à Rome à la Propagande, en 1846 [Par le P. J. Gagarin.] (Le Mans: Impr. de Julien, Lanier, Cosnard et Cie, 1858), 21 pp. *Любопытные свидетельства о непорочном зачатии богородицы*. Издал И. Гагарин ... (Le Mans, Imprimerie de Julien, Lanier, Conard et Cie, 1858), iv-11 pp.

27 Ioannikij Goljatovskij (died in 1688) — rector of the college founded in Kiev by Petr Mogila and later archimandrite of the Novgorod-Severskij Monastery. The "Ključ" refers to his collection of sermons published in Kiev in 1659, *Ključ razumenija*.

отмычка, с которою ходят слесаря и которою отпирают замки любой конструкции. Но что же далее? Вы правы: нет ни какого сомнения что в указанных Вами местах и Голятовский, и Барановский,[28] и Радивиловский[29] учили о непорочном зачатии в духе р[имско] католическом, но у них же мы встречаем места, полагаемые в основу веры, и духоборцами,[30] и штундистами...[31] Как о них рассуждать? Ведь это же не отцы Церкви и не вселенские учители. Мне жаль, что некоторые из них уже очень много наговорили. Среди польских проповедников XVI не то XVII в. был некто Млодзяновский,[32] который о таких неисследимых вопросах не спорил, но много говоря умел много учить народ тому, что его способно вести к истинному богопознанию и богопочитанию и я этого ксендза ставлю, как проповедника, гораздо выше приведенных Вами православных учителей, которые едва ли кого чему ни будь научили, а много напутали. Порою мне кажется, что некоторые из них сами не чуяли, что говорили, а говоря давали церковные концерты на своем красноречии. Это *казуисты*, а не веровдохновители. Вы конечно слыхали до чего это доходило, что студента, например, заставляли прежде доказывать известное положение, а потом отрицать его, и из этих-то искусников, из этих, так сказать религиозных гимнастов вышли те, чьи свидетельства Вы приводите, как бы нечто веское для совести людей нашей Церкви. Это именно *только любопытно*, — не более. Вам вероятно известно, что у этих писателей можно найдти свидетельства о том, что И[исус] Х[ристос] сын Божий как бы только *по духу*, и т.п.

К этому позвольте присоединить новость лично меня касающуюся: я кажется еще на месяц останусь в Париже и буду жить rue Chateaubriand 5. Все это произошло потому, что встретил приятелей, нашел комнату, удобную для работы и хочу попробовать кое-что сделать. — Если как ни будь про меня вздумаете, то напишите мне по этому адресу. Там же живет и Буслаев.[33]

[28] Baranovskij (*sic*) — Leskov obviously meant Lazar' Baranovič (1620-1694), archbishop of Černigov.

[29] Antonij Radivilovskij — prominent seventeenth-century Orthodox cleric, especially famous for his preaching, who eventually rose to be head of the Kievo-Pečersk Monastery.

[30] The Dukhobors — a Russian Protestant sect that arose in the Xar'kov region in the mid-eighteenth century. The persecution they suffered in the nineteenth century aroused the sympathy of Leo Tolstoj, and he played an important part in helping thousands of them to migrate to Canada at the turn of the century.

[31] Stundism forms one line of the spiritual ancestry of the present-day Russian Baptists. It arose in the early nineteenth century in the south of Russia when Orthodox Russian and Ukrainian peasants became converted to Protestantism through the influence of German settlers there. Leskov was greatly interested in the Stundists and wrote about them frequently. See William B. Edgerton, "Leskov, Paškov, the Stundists, and a Newly Discovered Letter", in *Orbis scriptus* (München: Wilhelm Fink Verlag, 1966), pp. 187-199.

[32] Tomasz Młodzianowski (1822-86) — Polish Jesuit famous for his sermons.

[33] According to Leskov's letter of June 23 [11] to his son Andrej (*Собрание сочинений*, X, 402-403), "Тут я буду жить с семьею моск[овского] профессора Буслаева, которым интересуется Коля [Nikolaj Mixajlovič Bubnov, Andrej Leskov's half-brother]; с г-м и г-жою Монтеверде, которых знает Ледаков, и с двумя сестрами Левиными, которые учатся в здешней консерватории и будут зимою петь на Мариинском театре. Так у

Какое исследование, или чье-то хождение куда-то издали Вы, или отец Мартынов, с своими примечаниями? Давно про это слыхал, а до сих пор ни как не добьюсь.[34]

Искренно уважающий Вас

Н. Лесков

The last letter in this correspondence is the reply to Leskov's fourth letter, quoted above, which Gagarin wrote on June 25/13 from Plombières.[35] After expressing pleasure over Leskov's plans to remain another month in Paris, which might give them an opportunity to see each other again, Gagarin went on to devote most of his letter to disputing what Leskov had said about the doctrine of the Immaculate Conception.

For all their importance, Leskov's associations in Paris with I. S. Gagarin were only one of the influences during his second trip abroad that led to his new orientation in religion. We know from a letter he wrote in Marienbad on August 10/July 29, 1875, to his friend P. K. Ščebal'skij that his beliefs had been shaken by a meeting with "young Naville". This is almost certainly a reference to Louis Naville, who was the second son of the eminent Swiss Protestant theologian Ernest Naville and who in 1872 had been made secretary of the École Libre des Sciences Politiques in Paris (now the Institut d'Études Politiques).[36] In this same letter to Ščebal'skij as well as in an article published in the following year Leskov wrote that while he was abroad he had read numerous books, both religious and political, that were forbidden in Russia.[37] His published writings

---

нас своя русская компания, и даже превеселая: мы вместе обедаем и вечером имеем даровую музыку и пение." Although Leskov stayed at this new address much less than the month he had planned (on July 15/3 he went to Marienbad), he retained very pleasant memories of his associations with the family of Buslaev (cf. his letter of June 1, 1877, to F. I. Buslaev, *Собрание сочинений*, X, 449-452), and in 1877 he dedicated his story *Некрещеный поп* to Buslaev.

[34] As Gagarin explained in his reply, the book about which Leskov inquired here was probably Meletij Smotrickij, *Apologia peregrinatiey do Kraiów wschodnych* (Lwów, 1628), reprinted in Leipzig in 1863 by I. M. Martynov.

[35] ЦГАЛИ, ф. 275, оп. 1, ед. хр. 218.

[36] *Собрание сочинений*, X, 411-412. Leskov's reference to "young Naville" could not have meant Ernest, the father, who by then was fifty-nine years old, nor could it have meant either of Louis Naville's brothers, Albert and Adrien. According to a letter of February 3, 1954, to me from Dr. François Naville, of Geneva, the grandson of Ernest Naville, both Albert and Adrien were living in Geneva in 1875. Moreover, Dr. Naville informed me that his uncle Louis read Russian fluently at the end of his life. Whether he knew Russian as early as 1875, when Leskov very likely met him in Paris, the Naville family does not know. Ernest Naville's interest in Russia and his numerous associations with Russians are described in the biography written by his granddaughter, Hélène Naville, *Ernest Naville, sa vie et sa pensée*, 2 vols. (Genève-Paris, 1913, 1917).

[37] Letter to P. K. Ščebal'skij of July 29 [August 10], 1875: "Более всего разладил с церковностью, по вопросам которой всласть начитался вещей, в Россию не допускаемых"

during the months following his return to Russia give us some indication of what these books were. In *Velikosvetskij raskol*, which began publication in the *Pravoslavnoe obozrenie* in September 1876, Leskov mentioned Ernest Naville no less than five times and quoted from two of Naville's books, *La Vie éternelle* (1861, Russian translation 1865) and *Le Problème du mal* (1868; published in Russian also). He likewise made five references to the Anglican theologian Canon Frederick William Farrar (1831-93); three to the great Swiss-American theologian Philip Schaff (1819-93), whose wide range of religious activities included working for closer relations among all branches of Christendom; and two references to "Gollard", whom he does not identify further but who is bound to be the French Protestant leader Roger Hollard, author of an *Essai sur le caractère de Jésus Christ* (1866), *La Jeunesse et l'Évangile* (1867), and *Méditations évangéliques* (Volume 1, 1874). It is clear that these four Protestant theologians were closely associated in Leskov's mind, for on two different occasions in *Velikosvetskij raskol* he refers to all four at once — "Naville, Hollard, Schaff, and Farrar" (on pp. 22 and 51 of the first edition in book form, Moscow, 1877). In 1876 he likewise mentioned for the first time Friedrich Rudolf Hasse (1808-62), the German professor of church history at the University of Bonn;[38] Ernest Naville's liberal Catholic friend Father August-Joseph-Alphonse Gratry (1805-72), who had been persecuted for his opposition to the new Catholic doctrine of Papal infallibility;[39] and the catechism of the Russian Molokans, *Veroispovedanie duxovnyx xristian, obyknovenno nazyvaemyx molokanami*, which was published in 1865 in Geneva by Herzen's Free Russian Press. This work, which circulated widely among the Dukhobors as well as the Molokans, had a great influence on Leskov. In his letter of August 10/July 29, 1875, to P. K. Ščebal'skij we can see a probable reflection of this book in his statement: "Had I read all the many things I have now read on this subject [the reunion of Christendom] and heard all that I have now heard, I would not have written *Soborjane* as I did write it, for that would have been distasteful to me. Instead, I am itching now to write about a Russian heretic — an intelligent, well-read, and free-thinking *Spiritual Christian* [the very name the Molokans and Dukhobors called themselves], who has passed through all doubts for the sake of his search for Christian truth and has found it only within his own soul. I would call this story 'Fornosov the Heretic', and I would publish it — but where would I publish it? Oh, these 'political tendencies'!"[40] As late as

(*Собрание сочинений*, X, 411); "Проведя все лето 1875 года за границею, я имел терпение перечитать там большое количество агитационных книжек ..." ("Энергическая бестактность", *Православное обозрение*, II, № 5 [май 1876], 133).

[38] "Сентиментальное благочестие", *Православное обозрение*, I, № 3 (1876), 540.

[39] "Дикие фантазии", *Православное обозрение*, III, № 11 (1877), 515.

[40] *Собрание сочинений*, X, 412.

1895, in his memoirs on Leskov, I. A. Šljapkin mentioned Leskov's study of the Molokan Catechism along with other details relating to the period immediately following his return from Western Europe in 1875.[41]

Leskov never wrote "Fornosov the Heretic", but he did write numerous other literary works that bore witness to the new direction his religious life had taken during his visit to Western Europe in 1875. Before the year was out he began publishing "Na kraju sveta",[42] which was far more heretical than Konstantin Pobedonoscev evidently realized when he recommended it to Tsar Alexander II a few months later for reading in his leisure moments.[43] In 1877 he published an even more heretical story, "Nekreščenyj pop", the hero of which was an Orthodox priest who had never been properly baptized and owed most of his Christian goodness to the influence of the Stundists.[44] In 1878 he began his series of sly, irreverent sketches about human frailties in the higher ranks of the Orthodox clergy, *Meloči arxierejskoj žizni*, which was banned from all public libraries in 1884 and was not republished thereafter in its complete form until 1957.[45] In 1879 he published another heretical story, "Odnodum", about a "Biblical socialist" who became acting mayor of a provincial town and threw the local economic life into turmoil by his refusal to accept bribes.[46]

The parallels between Leskov's and Leo Tolstoj's religious convictions brought the two men together in the mid-1880's; and Leskov remained a devoted, if not uncritical, follower of Tolstoy until the end of his life in 1895.

You know how much good you have done me [he wrote to Tolstoj on January 4, 1893]. From the earliest years of my life I was attracted to questions concerning religion and began writing about religious people when that was considered awkward and im-

[41] И. А. Шляпкин, "К биографии Н. С. Лескова", *Русская старина*, LXXXIV, № 12 (1895), 208.

[42] "На краю света. Рождественский рассказ", *Гражданин*, № 52 (1875); № 1-4, 6 (1876).

[43] Pobedonoscev's letter of May 14, 1876, to Alexander III: "Кроме того, позвольте еще представить вам книжку: *На краю света*, сочинение г. Лескова, — того самого, который написал известный вам рассказ: *Соборяне*. Книжка эта хорошо написана, и думаю, что она вам понравится, когда изволите прочесть ее в досужную минуту" (*Письма К. П. Победоносцева к Александру III*, I [M., 1925], 44).

[44] "Некрещеный поп (Легендарный случай). Посвящается Федору Ивановичу Буслаеву", *Гражданин*, № 23-24, 25-26, 27-29 (1877).

[45] "Мелочи архиерейской жизни. Картинки с натуры", *Новости*, № 236, 239, 240, 244, 248, 253, 259, 265, 272, 278, 285, 292, 298. Reprinted in book form in 1879 and 1880; reprinted in Н. С. Лесков, *Собрание сочинений*, VI (СПб., 1889), 245-510; but this edition was banned immediately after printing and eventually burned on order of the censors; reprinted in a form so severely censored as to be almost worthless in *Полное собрание сочинений Н. С. Лескова*, издание третье, XXXV-XXXVI (СПб., 1903); republished in its complete form for the first time since 1880 in *Собрание сочинений*, VI (M., 1957), 398-623.

[46] "Русские антики (Из рассказов о трех праведниках). Однодум (Очерк)", *Еженедельное Новое время*, III, № 37-38, 39 (20 и 27 сентября 1879), 641-660, 769-778; reprinted as "Однодум", *Собрание сочинений*, VI, 211-243.

possible (*Soborjane*, *Zapečatlennyj angel*, *Odnodum*, and *Meloči arxierejskoj žizni*, etc.); but I kept on with them and satisfied myself with the thought that I was "cleaning the rubbish out of the temple", but I did not know what to take into the temple ... when I heard your explanations, logical and powerful, I understood everything as if I had suddenly "remembered" it, and I no longer needed my own light, but began to live in the light that I saw coming from you, which was pleasanter for me because it was incomparably more powerful and brighter than the light in which I had been struggling on my own.[47]

The "humble heresiarch Nikolaj", as Leskov liked to sign himself in joking letters to his friends, continued to walk in the light from Jasnaja Poljana up to the very end of his earthly journey. The last known works to come from his pen, less than a month before his death, were articles devoted to Tolstoj and his forthcoming religious story "Master and Man" (*Xozjain i rabotnik*), which Leskov had the opportunity to read in proof shortly before it was published.[48]

[47] *Собрание сочинений*, XI, 519.
[48] "О повести Толстого", *Петербургская газета*, № 27 (28 января 1895). Andrej Leskov in *Жизнь Николая Лескова*, pp. 657-658, says that Leskov is definitely the author of this article and is probably also the author of another short note on Tolstoj's new story that appeared just the day before in *Новое время*, № 6794 (27 января 1895). Another article on Tolstoj that Leskov wrote only a few days before his death was not published until 1956: "О юбилее Л. Н. Толстого", *Огонек*, № 9 (26 февраля 1948).

THE COMIC ELEMENT IN *WAR AND PEACE*

MAURICE FRIEDBERG

Indiana University

In a recent study of stylistic devices of Tolstoj's novels a Soviet scholar pronounces a rather categorical verdict: "Humor is not characteristic of Tolstoj the novelist."[1] The Soviet writer Veresaev shares this view. In his opinion, Tolstoj is as humorless as children are — as if both of these assertions could not possibly be questioned.[2] The sentiment is echoed by yet another Soviet scholar, L. Plotkin, who asserts that Tolstoj merely assumes the pose of a "naïve outsider".[3]

War and Peace has traditionally been regarded as the epitome of the popular misconception of what a real novel of "high seriousness" ought to be — i.e., as a novel of unrelieved solemnity, written, as it were, in accordance with Aristotle's and Boileau's injunctions against the breaking down of the walls traditionally separating Thalia and Melpomene in ancient and neo-classical drama. The fact that long literary works cannot, as a rule, support a huge weight of sobriety without some sort of comic relief has somehow been overlooked in the numerous investigations of Tolstoj's art, in spite of the fact that even the ancients of Aristotle's day were quite aware of this limitation: while the written *texts* of Greek tragedies may have shunned the comic, *performances* of the very same tragedies were intentionally and regularly interrupted with comic fragments, clowning, etc., in order to provide the spectator with some respite from the solemnity of the drama, before the advent of the final or "true" catharsis which was to be achieved at the end of the play. It appears to this writer that the massive body of narration in *War and Peace* is made more easily digestible to the ordinary reader — perhaps the Russian term *udobočitaemyj* might express it better — by the often maligned philosophical and historical digressions, which are normally the first to be sacrificed in any abridged version of the novel, thus defeating the original purpose of the abridgement. One of the functions of these digressions is to relieve the tedium of hundreds of pages

[1] A. Čičerin, "Stil romanov L'va Tolstogo", *Russkaja literatura*, No. I (1963), p. 44.
[2] V. V. Veresaev, *Sočinenija v četyrex tomax*, II (Moskva, 1947), p. 554.
[3] L. Plotkin, *Literaturnye očerki i stat'i* (Leningrad: Sovetskij pisatel', 1958), p. 253.

of descriptions of happy families — which, as Tolstoj himself admits in the opening lines of *Anna Karenina*, are, in the final analysis, all alike — and of the thousands of petty details of everyday life. The stern, impersonal quality of these pages of theorizing on the nature of Man and of War and of History is not unlike that of the chorus in the ancient tragedy. It breaks the threat of monotony which is always inherent in a prolonged account of necessarily trivial events in the lives of ordinary humans. In fact, these digressions are, in this writer's opinion, to be counted among the important stylistic devices which set apart *War and Peace* from its lesser "epic" brethren, such as Margaret Mitchell's *Gone with the Wind.* Indeed, it is not unlikely that the contemporary American novelist Herman Wouk introduced lengthy excerpts from the United States Navy regulations into his *Caine Mutiny* with a similar purpose in mind.

Less controversial is the cathartic function of the estrangement (*ostranenie*) device. Yet Viktor Šklovskij in his famous study of the style of *War and Peace*, preoccupied as he is with the twin problems of the estrangement and the fictionalization of historical material, fails, it seems, to appreciate the fact that the very device of estrangement owes much of its effectiveness to the profoundly comic flavor of the *literal* meaning of scenes presented in this manner. Surprising as this may seem, Šklovskij's treatment of the comic element in the novel is all but exhausted by the remarks that Tolstoj frequently resorts to the use of German in order to achieve a comic effect,[4] and to the rather sententious dictum that "the lines of *War and Peace* are permeated with cold, horsey irony".[5]

The insufficient attention devoted to the subject is perhaps also due in part to the overall neglect of the problem of the comic in investigations of Russian literature, aggravated by the fact that there are only four works of Tolstoj in which the comic element may be said to occupy a central position, namely, the plays *The First Distiller*, *Fruits of Enlightenment* and *The Contaminated Family*, as well as his tract *Nikolaj Palkin.* Be that as it may, there has been, to this writer's knowledge no overall study of the comic element in Tolstoj comparable in scope to Slonimskij's classic monograph on Gogol',[6] although there have been several shorter studies devoted to individual works.[7] Generally, Soviet investigators emphasize Tolstoj's satire and tend to confuse the latter with the

[4] Viktor Šklovskij, *Material i stil' v romane L'va Tolstogo "Vojna i mir"* (Moskva: Federacija, 1928), p. 214.
[5] *Op. cit.*, p. 28.
[6] Aleksandr Slonimskij, *Texnika komičeskogo u Gogolja* (Petrograd: Akademija, 1923).
[7] E.g., V. Čalmaev, "O nekotoryx èlementax komedijnogo stilja v *Plodax prosveščenija*", *Lev Nikolaevič Tolstoj: sbornik statej o tvorčestve*, pod obščej redakcijej prof. N. K. Gudzija (Izd. Moskovskogo universiteta, 1959), pp. 113-30; L. Kuzina, "Razvitie satiry v tvorčestve L. N. Tolstogo pozdnego perioda (s 80-x gg. XIX v.)", *ibid.*, pp. 131-58.

comic in general. While one cannot quarrel with the view that satire plays an important role in several of Tolstoj's works, the indiscriminate identification of satire with the comic seems unjustified, and Edgar Johnson appears more convincing when in the introduction to his anthology of satire he

... insists that satire need not be comic and cites to prove his contention, in addition to Juvenal's direct invective (which is not ironical), the irony of Swift on the Yahoos and Aldous Huxley's on the Fifth Earl of Gonister.[8]

The thesis of the present paper is that while Tolstoj is no humorist in the sense that Gogol' is, i.e., the comic to Tolstoj is more often a means toward an end than an end in itself, Tolstoj was deeply aware of the artistic potential of the comic and, in *War and Peace*, consciously employed a great variety of comic devices. Even toward the end of his life the "grim" writer was not insensitive to the attractions of the comic. In his letter to V. G. Čertkov of January 15, 1890, Tolstoj wrote:

A man is about to die any day now, but all of a sudden he carefully jots down a sentence which fits the personality of a particular character *and which is funny*. And the man is overjoyed at finding it.[9]

There are many definitions of the comic, but the one which is perhaps most readily applicable to Tolstoj's *War and Peace* is the well-known formula of Henri Bergson, who described the comic as "something mechanical encrusted upon the living".

The mechanical is, indeed, shown encrusted upon the living in the opening scene of the novel. At the "informal" party of Anna Pavlovna Scherer, a lady-in-waiting of the Empress, we first meet Prince Vasilij. He is shown talking "from habit, like a wound-up clock, saying things he did not even wish to be believed".[10] As the room filled,

All the guests performed the ritual of greeting [Anna Pavlovna's senile] aunt, whom nobody knew, or found interesting or necessary. *Ma tante* uttered to each guest the same words about his health, her own health, and the health of Her Majesty, who was, thank God, better of late. Every one, though from politeness showing no undue haste, moved away from the old lady relieved at having discharged a burdensome duty, and did not approach her again all evening. (I, 11-12)[11]

[8] Alan Reynolds Thompson, *The Dry Mock: A Study of Irony In Drama* (Berkeley and Los Angeles: University of California Press, 1948), p. 12.

[9] Quoted in Čalmaev, *op. cit.*, p. 114. Italics supplied.

[10] L. N. Tolstoj, *Vojna i mir*, posleslovie S. Byčkova (Goslitizdat, 1949), I, 6. The text of this two-volume edition of the novel is identical with that of the Jubilee Edition of the complete works of the novelist. All subsequent references will be to the two-volume set, the Roman numeral indicating the volume and the Arabic the page.

[11] The description bears a striking similarity to a scene found in Puškin's *Queen of Spades*, in which the old countess is the object of the same ritual at a ball, and, to a lesser extent, to the famous opening stanza of *Eugene Onegin*.

The "spontaneity" of conversation at the gathering is mercilessly mocked in the lines where Anna Pavlovna, the hostess, is compared to an owner of a spinning-mill who "noting any stoppage or unusual creaking or too loud a whir in the spindles, goes up hurriedly, slackens the machinery and sets it going properly". Anna Pavlovna, "walking about her drawing room, went up to any circle that became silent or overly animated and by a simple word or change of position set the conversational machine going again in its regular, proper manner" (I, 14). Indeed, her greatest worry is caused by Pierre and the *Abbé*: "both were listening and talking too eagerly and too naturally"! (I, 18).

Berg and Vera, after their marriage, are also "entertaining", i.e., they have invited a rigidly selected group of people to spend an evening doing strictly prescribed things and chatting about narrowly limited subjects in an agreed-upon tone of simulated informality: "the *soirée* was launched very well, and the *soirée* was a carbon copy of every other *soirée*, complete with conversation, tea and burning candles" (I, 557-58).[12]

The strict rules governing the flow of ostensibly informal conversation in high society are also shown in the scene where, during Anatole Kuragin's disastrous visit to the Bolkonskij estate, upon hearing the word "Paris", Mademoiselle Bourienne, as if responding to a conditioned reflex, decides that this is the proper time for her to join in the discussion (I, 266).

The artificiality of the society gathering is underlined by Tolstoj's repeated usage of "gastronomical" similes and metaphors, comical in their unexpectedness. Thus, at the beginning of the novel, a certain *vicomte* is "served to the guests":

> As a resourceful *maître d'hôtel* serves as something extraordinarily wondrous the piece of beef which no one would eat had he found it in a dirty kitchen, so did Anna Pavlovna serve on that evening to her guests first the *vicomte* and then the *abbé* as a most sophisticated delicacy (I, 15).

She was to do it again four hundred pages later in the novel (I, 433).[13]

The ridiculous results of the hypocrisy enveloping society gatherings — or, as

[12] To strengthen the impression of pretense, the newly-married smiling hosts are revealed to hold each other in contempt. Berg believes that Vera, being a woman, is, of course, his intellectual inferior. Vera, on the other hand, is convinced that her husband, a mere male, "thinks that he alone is clever; but in reality he understands nothing and is conceited and selfish" (I, 557-58). Berg is also said to call a mere acquaintance a friend "because he knew that all people have friends" (I, 531).

[13] A "cannibalistic" association of another rite, this time a religious one, is found in the notorious scene of Church service in Tolstoj's *Resurrection* in which the participants are shown enjoying eating "Christ's flesh" and drinking "Christ's blood". Needless to say, the passage was found offensive by many Christians.

"modern" critics might say, of their loss of innocence — are pungently illustrated by Tolstoj's remark that on one occasion Mademoiselle Bourienne was "equally ready to break into tears or to burst out laughing" (I, 117); by the impression made by Hippolyte at a party at which all the guests were left stunned and unable to decide whether he is very stupid or very profound (I, 17); and, most eloquently, by the fact that the essentially shallow and vulgar Ellen, for a time Pierre's wife, gradually acquires a reputation as an unusually intelligent and tactful woman by the simple stratagem of uttering meaningless remarks alternated with periods of silence.

Conventionalized social relations are caricatured by Tolstoj in his descriptions of insincerely sentimental courtships and hypocrisy in marital relations, subjects which are also treated at length in many of Tolstoj's other works. Thus, on one occasion, "Berg, a tender smile on his face, was discussing with Vera the fact that love is not an earthly but a heavenly feeling" (I, 75). During Anatole Kuragin's visit to the Bolkonskijs, Mademoiselle Bourienne daydreams of a prince charming — here, obviously, Anatole — who would seduce her, then for a time live with her in sin, but later, after seeing the unhappiness this would cause to Mademoiselle Bourienne's *pauvre mère* (who would reproach her daughter for her immoral life), would marry her (I, 269-70).[14] Tolstoj then proceeds with the cynical observation that Mademoiselle Bourienne and Anatole seem to have reached complete understanding "concerning the first part of the romance, i.e., that preceding the appearance of *pauvre mère*" (I, 273).[15]

There is much sarcasm in the account of Boris Drubeckoj's courtship of the wealthy spinster Julie. While it is clear to everybody that Boris is only interested in her money, while she *must* marry before getting too old, the two pretend for a long time to be nothing more than kindred souls comforting each other in a cruel, insensitive world.

[14] This is obviously a lampoon of one of the favorite themes of sentimentalist literature — the seduction of a simple maiden by a nobleman — a plot made famous by Karamzin's *Poor Liza* (a novel which provokes Tolstoj's mocking comment elsewhere in the novel, namely, in the scene describing the courtship of Julie and Boris, I, 653-54), and then half-seriously re-told by Puškin in "The Stationmaster", one of *The Tales of Belkin*, where the "happy" ending (the noble seducer marries his innocent victim) is balanced by a "tragic" postscript: the girl's father dies from grief without ever learning that his daughter's dishonor was erased by her marriage.

[15] Tolstoj is very amused at the sentimental invocation of the memory of one's mother — *ma pauvre mère* — which he defines as a national characteristic of the French. Thus, e.g., before telling Pierre the odyssey of his amorous exploits, the French captain Ramball who took Pierre prisoner in occupied Moscow, speaks of *ma pauvre mère* (II, 367). Napoleon himself "like all Frenchmen, could not imagine anything touching without recalling *ma chère, ma tendre, ma pauvre mère*" (II, 321), and orders that all hospitals, orphanages and similar institutions in French-occupied Moscow bear the inscription *Maison de ma Mère*. To the great amusement of the Russians, this dedication was also engraved on the city's insane asylum.

Boris drew in her album two trees and wrote: *Arbres rustiques, vos sombres rameaux secouent sur moi les ténèbres et la mélancolie.*

Elsewhere he drew a tomb and wrote:

La mort est secourable et la mort est tranquille
Ah! contre les douleurs il n'y a pas d'autre asile.

Julie said that this was delightful.

— *Il y a quelquechose de si ravissant dans le sourire de la mélancolie,* — she repeated to Boris word by word a passage she had copied from a book.

Whereupon Boris wrote her the following verses:

Aliment de poison d'une âme trop sensible,
Toi, sans qui le bonheur me serait impossible,
Tendre mélancolie, ah, viens me consoler,
Viens calmer les tourments de ma sombre retraite
Et mêle une douceur secrète
A ces pleurs, que je sens couler.

Julie played for Boris on her harp the saddest nocturns. Boris read to her aloud [Karamzin's sentimental novel] *Poor Liza,* stopping more than once from agitation which interfered with his breathing. (I, 653-54)

In the end, Boris proposed to Julie not a mystical union of kindred souls, but an ordinary marriage:

"You know my feelings for you!"

It was not necessary to say more. Julie's face radiated truimph and smugness. Nevertheless, she compelled Boris to tell her all that is normally said on such occasions, to tell her that he loved her and that he had never loved another woman more than he loved her. She knew that her Penza estates and Nižnij Novgorod forests entitled her to demand it, and she received what she demanded.

The bride and groom spoke no more of trees covering them with darkness and melancholy, they were engaged in planning the arrangements for a resplendent house in Petersburg and preparing for a dazzling wedding. (I, 656)

Even more effective is a passage describing Pierre's proposal to Ellen:

"There is something special that people say on such occasions", he thought, but could not possibly recall what precisely people *do* say on such occasions ... "*Je vous aime!*" he said, recalling what must be said on such occasions, but the words sounded so hollow that he felt embarrassed. (I, 255)

Shortly before Pierre's proposal, the bride's father, Prince Vasilij, is shown exchanging platitudes with a dowager:

— Конечно, *c'est un parti très brilliant, mais le bonheur, ma chère...*
— *Les mariages se font dans les cieux,* — отвечала пожилая дама. (I, 254)

It is interesting to note that the very same platitude — *les mariages se font dans les cieux* — is reported without a trace of irony when uttered by Nataša Rostova's mother with reference to Andrej Bolkonskij's proposal to her daughter (I, 563). Nor was there any irony in Tolstoj's most famous scene of marriage proposal, that of Levin and Kitty in *Anna Karenina*, a description reportedly based on Tolstoj's own proposal to Sophie Behrs. Awkwardness and cliché, the staple excuses for comedy, are used by Tolstoj to create sympathy for his characters, provided these characters are honest, upright, etc. Tolstoj never laughs at real, and particularly at undeserved, misfortunes; his laughter is directed at pretense, pomposity, at the "mechanical". That is why no reader would dream of smiling at the ridiculously pathetic attempts of Princess Marie to make herself pretty on the occasion of Anatole Kuragin's visit to their estate (I, 261) or at the little moustache of Lise Bolkonskaja (e.g., I, 12: "её хорошенькая с чуть черневшимися усиками верхняя губка"). Tolstoj's notion of the comic is definitely tempered by ethical considerations.

Perhaps the best known instances of the comic in the novel are to be found in Tolstoj's philippics against the striking — and hence, comic — incogruities of war, particularly against the notion that wars may be planned, directed, and that there may even be a military science. The full ferocity of Tolstoj's ridicule is directed at those of the military who take themselves very seriously, be they French or Russian.

Thus, we are informed that the Russian general staff was, ethnically, anything but Russian. In fact, its deliberations were hindered by formidable language barriers:

> Paulucci, who did not know German, began to ask him questions in French. Woltzogen came to the assistance of his leader who spoke French very badly, and began translating his utterances ... Paulucci and Michaud fell simultaneously on Woltzogen in French. Armfeldt addressed Pfuhl himself in German. Toll was interpreting to Prince Bolkonskij in Russian. Prince Andrej listened and watched them in silence. (II, 52)

An extremely funny scene is created by a torrent of patriotic Russian sentiments sounded at the height of Franco-Russian hostilities during a conversation between the Russian Emperor Alexander and one of his officers, a man who "expressed his own feelings as well as those of the Russian people whose spokesman he considered himself to be". The officer's name was Michaud; the officer was addicted to describing himself as "*quoique étranger, mais russe de cœur et d'âme*". Naturally, the Russian Emperor and the Russian officer conversed in ... French (II, 404-05).[16]

[16] As the wave of Russian patriotism spreads in Russian high society, Russian noblewomen

A devastating caricature of a military figurehead who is taking himself very seriously is found at the beginning of the second volume of the novel:

They had just passed an inn and were riding uphill when a party of horsemen came riding downhill towards them, lead by a tall man, in a hat with plumes, on a black horse, with trappings glittering in the sun. He had curly black hair that floated on his shoulders, a scarlet cloak, and he rode in the French fashion, with his long legs thrust out in front. This man galloped toward Balašov, with his jewels, gold lace and feathers all fluttering and glittering in the bright June sun.

Balašov was some ten yards from this majestically theatrical rider in bracelets, feathers, necklace, and gold, when Julner, the French colonel, whispered to him reverentially, "*Le roi de Nâples*". It was in fact Murat, who was now called the King of Naples. Though it made no sense why he should be the King of Naples, he was addressed by that title, was himself convinced of that fact, and consequently assumed an air of greater solemnity and dignity than he had heretofore. He was so firmly convinced that he really was the King of Naples, that when, just before leaving Naples, whiie taking a walk in the streets with his wife, he was greeted by several Italians with shouts "*Viva il re!*", he turned to her with a sad smile and said, "*Les malheureux, ils ne savent pas que je les quitte demain.*" (II, 19)[17]

The tragicomedy of pretense is even more evident during the disorderly retreat of the French Army:

All were marching, not knowing where they were going or why. Napoleon, the genius, knew it even less than the others, since nobody was giving *him* any orders. Nevertheless, he and his entourage observed their old customs. They addressed each other *Sire, Mon Cousin, Prince d'Ekmuhl, Roi de Nâples*, etc. (II, 553-54)

To be sure, Tolstoj finds similar incongruities among the Russians. Thus, after an *unnecessary* battle which cost many human lives, Kutuzov "received a diamond decoration. Benningsen also received diamonds and 100,000 rubles. Others also received many nice things, each according to his rank" (II, 473). Before an earlier battle, Kutuzov, having listened to many learned and clever expositions of military strategy, dismisses the meeting with the remark that it is too late to change anything: "And before a battle nothing is more important ... (he stopped for a moment) than getting a good night's sleep" (I, 314).

---

invent a parlor game: anyone caught speaking French must pay a fine. The following conversation ensues:

— Вы знаете, что я в самом деле думаю что она *un petit peu amoureuse du jeune homme.*

— Штраф! Штраф! Штраф!

— Но как же это по-русски сказать! (II, 177)

The last remark brings to mind Puškin's tongue-in-cheek remark in *Eugene Onegin*; after using several foreign words (to be sure, on purpose), Puškin says, plaintively: "Шишков, прости: не знаю как перевести."

[17] The incident bears some similarity to the apocryphal story according to which Catherine the Great persistently refused to believe there might be any Russian peasants who opposed her, basing her conviction on the fact that the peasants had always called her *matuška.*

Elsewhere, Tolstoj devotes several pages to an intentionally detailed description of the various factions within the Russian general staff (II, 42-46). The fact that Tolstoj discerns *eight* factions (rather than two or three) serves to underline the impossibility of serious military planning, the pettiness of the various groupings, their jealousies, biases, etc., as well as the fact that the seemingly irreconcilable factions do not, *in essence*, differ very much from each other — a comic device used by Jonathan Swift in *Gulliver's Travels*, particularly in the account of the Lilliput parties quarrelling about which end an egg should be broken on, and also found in Russian literature, e.g., in the "fairy tales" of Mixail Saltykov-Ščedrin, such as his mock-serious theorizing on the "basic" differences dividing "moderate" liberals and "enlightened" conservatives in nineteenth-century Russia.

The comic effect in *War and Peace* is sometimes heightened by ironically-solemn pseudo-scientific classification of events and ideas (which, indeed, facilitates their acceptance as "serious"), a device used most advantageously in the short, epigrammatic enumeration of the many national varieties of conceit (самоуверенность):

Pfuhl was one of those hopelessly, immutably conceited men, ready to face martyrdom for their own ideas, conceited as only Germans can be, just because it is only a German's conceit that is based on an abstract idea — science, that is, the alleged possession of absolute truth. The Frenchman is conceited from believing his person to be irresistibly bewitching to men as well as to women. The Englishman is conceited on the ground of being a citizen of the best constituted state on earth, and also because, being English, he always knows what is the correct thing to do, and knows that everything that he, as an Englishman, does, is indubitably good. The Italian is conceited from being temperamental and easily forgets himself as well as others. A Russian is conceited precisely because he knows nothing, and does not want to know anything, believing it impossible to know anything fully. A conceited German is the worst of them all, and the most hardened of all, and the most repulsive of all, because he imagines that he possesses the truth in a science which he himself invented, but which, he believes, is the absolute truth. (II, 49)

The passage is also comic in the extreme because the dogmatically generalized judgements of Tolstoj are, of course, in themselves either scathing self-irony or else glaring manifestations of the author's own monstrous conceit.[18]

In his later works, i.e., those written during his "pacifist" period, Tolstoj would frequently proclaim in thundering tones the irreconcilability of Christianity with the waging of war under any circumstances. Foreshadowings of this

[18] Other generalizations are less obviously comic. Thus, e.g., Tolstoj's observation that Anna Pavlovna "paused, as women always do after mentioning their age, as if anticipating something" (I, 246) may or may not strike the reader as funny, depending, of course, on the reader's opinion of women. The present reader is willing to take the risk of going on record as finding the remark both true and amusing.

attitude are evident, however, even in *War and Peace*, which was written long before Tolstoj's "conversion", except that here Tolstoj is still capable of retaining a degree of detachment, manifested in the humorous tone of the two central remarks pointing out the awkward position in which the Church is placed during a war between two Christian nations. Thus, in one passage, Pierre is bewildered by the fact that

The Spaniards exalt the Lord through the intermediary of the Roman Catholic clergy in gratitude for their victory over the French on June 14, while the French exalt the Lord through the intermediary of that very same Roman Catholic clergy in gratitude for *their* June 14 victory over the Spaniards. (I, 639)

Elsewhere we observe the confusion of Nataša Rostova who, being a good Christian, must pray *for* the French, but, being a good Russian, simultaneously prays for the victory of Russia over France, i.e., in effect, prays *against* the French (II, 71).

War and Peace is replete with Tolstoj's acid comments about various generally accepted historical accounts of the Napoleonic campaings, in particular the monumental study of Adolphe Thiers (1797-1877), whom Tolstoj wishes to discredit by all possible means, including distorted summaries of his writings.[19] Occasionally, however, in an effort to destroy the arguments of his opponents, Tolstoj resorts to ridicule, and one cannot deny the devastating effectiveness of a passage like the following:

Louis XIV was a very haughty and self-centered man; he had such and such mistresses, and such and such ministers, and he governed France badly. Louis' successors were also weak men and they also governed France badly. And they had such and such favorites, and such and such mistresses. Furthermore, certain persons at that time were writing books. At the end of the eighteenth century some twenty persons gathered in Paris and they began to talk all about men's being equal and free. As a result, all over France people began to drown each other and cut each other's throats. These people killed the king and many others as well. At that very same time there was in France a man of genius, Napoleon. He conquered everyone everywhere, i.e., he killed a lot of people, because he was very much a genius. And for some reason or other he went to kill the Africans, and he killed them so well, and he was so cunning and clever, that upon his return to France he ordered everybody to obey him. And everybody obeyed him. Having made himself Emperor he went off to kill people in Italy, Austria, and Prussia. And there, too, he killed a great many. As for Russia, there was an Emperor there, Alexander, who decided to restore order in Europe and for this

[19] Thus, as Šklovskij points out, Tolstoj's claim that Napoleon resorted to the forging of Russian currency — information allegedly obtained from Thiers — is not only nowhere to be found in the writings of the French historian, but was not even advanced by the Russians themselves during the Napoleonic wars when they would have, understandably, been only too eager to point out this instance of "ungentlemanly" tactics on the part of their adversaries. See Šklovskij, *op. cit.*, pp. 177-78.

reason made war with Napoleon. But in 1807 he suddenly became Napoleon's friend, while in 1811 they quarrelled again and once again they began to kill off a lot of people. Then Napoleon took six hundred thousand men into Russia and he conquered Moscow, and then he suddenly ran away from Moscow, and then the Emperor Alexander, aided by the advice received from Stein and others, united Europe for defense against the destroyer of her peace. All of Napoleon's allies suddenly became his enemies. The allied forces advanced against the fresh troops assembled by Napoleon. The allies vanquished Napoleon, entered Paris, forced Napoleon to abdicate, and sent him off to the island of Elba, not depriving him, however, of his imperial title and treating him most respectfully, even though only five years before and one year later everybody considered him an outlaw and a bandit. And Louis XVIII, who until then had been a laughing-stock to the French as well as to the allies, began to reign. As for Napoleon, he shed tears before the old guard, abdicated the throne and went in exile. Later on subtle statesmen and diplomats (particularly Talleyrand, who succeeded in sitting down in a well-known chair before anyone else and thus extended the frontiers of France) conversed with each other in Vienna, and these conversations made some nations happy and others unhappy. And then diplomats and monarchs came to the verge of quarrelling. They were just about ready to order their armies to kill each other once more. But at that time Napoleon entered France with a batallion, and the French, who had been hating him, immediately all submitted to him. This made the allied monarchs angry, and again they went to war with the French. And Napoleon, the genius, was defeated; and suddenly recognizing that he was a bandit, they took him to the island of St. Helena. And there the exile, separated from his dear ones and from his beloved France, died on a rock a slow death and bequeathed all his great exploits to posterity. And in Europe a reaction took place, and all the sovereigns began oppressing their peoples once again.

Tolstoj continues:

It would be quite wrong to assume that all this is mockery, a caricature of historical descriptions... What is strange and comic in these answers is due to the fact that modern history is like a deaf man answering questions which were not even asked of him (II, 688-90).

The comic quality of the passage is caused by the overuse of simplification. All the customary qualifying statements, all the usual elaboration were dispensed with. As a result, the reader faces an enormous amount of historical material compressed into very small space. Whenever this becomes necessary, the normal practice is to retain only generalizations; Tolstoj does the very opposite: omitting generalizations, he retains only details. The effect is approximately the same as that of playing a 33 rpm. recording at the speed of 78 rpm. By increasing the speed of delivery without a corresponding abridgement in the quantity of sounds, one achieves the effect of gay and totally nonsensical chatter used in the sound tracks of animated film cartoons, such as Donald Duck or Mickey Mouse. The impact is heightened by Tolstoj's usage of very primitive, almost childish, language to describe momentous historical events. To clarify the point, let us "rewrite" the first part of the passage, without adding or deleting any of the essential information contained in it:

The period of the reign of Louis XIV was characterized by such corruption at the court and incompetence of administration, that, in spite of the monarch's towering personality, it must be classified as one of the more unfortunate periods in French history. In essence, things have not changed appreciably under Louis XIV's successors. The new administration continued the policies of the preceding one, and the mores of the court itself, historians agree, continued to shock even the most tolerant of observers. Nevertheless, the period will be remembered as one of a true cultural renaissance. While for the time being, of course, restricted to a small intellectual elite, the ideas of human equality and liberty gradually began to gain acceptance, though, to be sure, not without much controversy and, on occasion, violence.

The above re-telling of the corresponding passage from Tolstoj is no longer comical because we have deleted some elements of detail that were speeding up the narrative ("such and such mistresses and such and such ministers," "some twenty persons") and replaced them with "braking", slow-moving qualifications and generalizations; second, because we added some "meaningless" information in order to slow down the pace of the narrative even further ("historians agree", "most tolerant of observers"); and, finally, because Tolstoj's artificially naïve vocabulary was replaced by a "sophisticated" one, i.e., one which we have grown accustomed to consider appropriate in accounts of important events in human history.[20]

Quite frequently, Tolstoj provokes laughter by applying the seemingly opposite device of periphrasis or circumlocution, i.e., "using too many words to express something which could be put more briefly", a technique used, for example, by Alexander Pope in *The Rape of the Lock*.[21] Most often, Tolstoj merely repeats an ironical description of an individual instead of simply giving that person's name: "*C'est le doute qui est flatteur*, сказал *l'homme à l'esprit profond* с тонкою улыбкой" (I, 435), or "9-ого августа князь Василий встретился опять у Анны Павловны с *l'homme de beaucoup de mérite*" (II, 129).

[20] The speed of narration is also one of the chief comic devices in Gogol'. There is, however, an important difference: whereas Tolstoj's narrative in the passage just cited is, necessarily, hurried because the author has chosen to compress a very long story into a single page, Gogol''s narrative must move very rapidly because his "essential" story is crowded by numerous "superfluous" digressions. In order to relate the "essential" events at a "normal" pace, Gogol' is forced to increase the tempo of the entire narrative. *The Overcoat*, for example, lapses into a digression of some fifteen lines immediately after the opening words "in the Bureau of", and the fifteen lines of "irrelevant" material must be disposed of hurriedly in order to reach the next "essential" words — "a certain Bureau". A similar comic technique is employed by the Czech writer Jaroslav Hašek in *The Good Soldier Schweik*, in which the protagonist, extremely fond of storytelling, must always talk very rapidly in order to be able to squeeze into the allotted time both the "businesslike" message and the "irrelevant" tale.

[21] Karl Beckson and Arthur Ganz, *A Reader's Guide to Literary Terms* (New York: Noonday Press, 1960), p. 153.

The most extensive use of the periphrasis is to be found, of course, in the famous scene at the opera where Tolstoj employs what has since become known as остранение:

The stage had in its center a floor of evenly cut boards. On the sides there were sections of painted cardboard which represented trees. In the back there was linen cloth stretched over some boards. Maidens clad in red bodices and white skirts were sitting in the center of the stage. One of them, very fat, wearing a white silk dress, was sitting apart on a low bench with a green cardboard glued to its back. All of them were singing something. When the song was finished, the maiden dressed in white moved toward the prompter's box, and then a man with his fat legs squeezed in silk tights, with a plume and a dagger, approached her and began singing and waving his arms.

The man in the tights sang a piece alone, then she sang hers alone. Then both of them stopped while the music played, and the man fumbled with the hand of the woman in white, obviously waiting for the cue at which he was to begin to sing his part together with her. They sang together, and every one in the theater began clapping and shouting, and the man and the woman on the stage, who were playing the parts of lovers, began to bow, smile and wave their arms. (I, 666)[22]

In the second act there were pictures representing monuments, and there was a hole in the linen cloth which represented the moon, and shades were put over the footlights, and trumpets and bassoons began playing, and many people wearing black cloaks came in from the right and from the left. The people began to wave their arms, and in their hands they all had something resembling daggers. Then still other people ran in and began dragging away the maiden who had been wearing before a white dress but was now wearing a light-blue one. They did not drag her away immediately. First they sang with her for a long time. Then they did finally drag her away and behind the scenes they struck something metallic three times, and then they all fell on their knees and began chanting a prayer. All these actions were periodically interrupted by enthusiastic shouts from the audience. (I, 669)

In the third act the stage represented a palace with many candles burning and with pictures showing goateed knights. The pair standing in the center were, probably, the king and the queen. The king began waving his right hand and, obviously nervous, sang something very poorly and sat down on a purple throne. The maiden who had at first been wearing white and then light-blue, had on her now nothing but a chemise, and stood near the throne with her hair hanging loose. Addressing the queen, she was singing mournfully about something or other. But the king waved his hand sternly, and then from the sides there filed in bare-legged men and bare-legged women and they all began to dance together. Then the violins played very shrilly and gaily, and one of the maidens with fat bare legs and thin arms, leaving the others, went behind the curtain, adjusted her bodice, then walked into the center of the stage and began to jump and to kick rapidly one leg with the other. Everyone in the orchestra seats clapped their hands and shouted "bravo!" Then one man stood alone at one corner of the stage, the cymbals and trumpets struck up more loudly in the orchestra, and that very same bare-legged man began jumping very high in the air and rapidly waving his legs (this

[22] Constance Garnett in her translation of the novel renders "они пропели вдвоем" as "they sang a duet" which, of course, destroys Tolstoj's intentionally "naïve" language.

was Duport, whose remuneration for this art was sixty thousand a year). (II, 670)[23]

The periphrasis in the three passages cited above consists (as was pointed out in Šklovskij's discussion of остранение) in presenting the scene as it would be seen through the eyes of one who does not know or, as is the case here, has temporarily forgotten the conventions of the make-believe devices of the opera. What we want to stress is that the technique is successful precisely because it injects a comic note into the narrative. The reader cannot but be amused both by the "naïveté" of Nataša, who does not seem to know anything about such commonplace things as the opera and therefore uses a child's vocabulary to describe it, *and* — paradoxical as this may seem — *at the same time* pity and mock those gullible enough to take this make-believe seriously. Finally, the beautiful mirage of the opera is made to appear ridiculous by the reminder that the subtle art of leaping into the air carries the enormous salary of sixty thousand rubles: "pure" art is thus consciously soiled by the exposure of its intimate connection with monetary gain. The soap-bubble of romantic pretense is punctured, and this cannot but evoke laughter on the part of the reader, who is eager thus to create a barrier between himself and the silly believers in the pretenses of the opera.[24]

There are only a few instances in *War and Peace* of "conventional" comic devices, such as slapstick or situation comedy. Thus, Nataša quotes Pierre as saying that his hand was purposely made to fit the shape of their infant's posterior (II, 663). An unexpected "animal" metaphor is used when a middle-aged lady is described as walking иноходью, i.e., pacing or ambling (I, 93). More involved is the comical scene during the already mentioned visit by Anatole to the Bolkonskij estate: Anatole's official purpose is to propose to Princess Marie, but while going through the necessary motions, he simultaneously flirts with Prince Andrej's wife and at the same time also tries to seduce Mademoiselle Bourienne.[25]

If some — though not many — of the comic passages cited above may be salvaged in translation, there is at least one important category which is inaccessible to those who do not read the novel in the original. These are the instances

[23] Same as above. Garnett "corrects" Tolstoj's "simple-minded" девица ("maiden") by translating it into English by means of the "sophisticated" term "actress" (*ibid.*, p. 527).

[24] The technique itself was not, of course, invented by Tolstoj. Similar devices may be found in Swift's *Gulliver's Travels* and even in folkore of various peoples. Thus, e.g., in Martin Buber's anthologies of old Hassidic lore there is a tale of a deaf man who is bewildered by the sight of a wedding celebration, which to him appears as a silly scene of several men hitting with their fingers pieces of skin tightly stretched over barrels or blowing air into metal tubes, while other men and women twist their bodies as if they were in pain and walk around in intricate little steps.

[25] The situation is reminiscent of Gogol''s *Inspector General*, in which Xlestakov simultaneously courts the Mayor's wife and their daughter.

of comic effect achieved by using languages other than Russian in conjunction with Russian. Some of the situations are quite simple and traditional — e.g., a German mispronounces the sounds of Russian (I, 85).[26] A similar technique is used in a scene depicting a Russian soldier's attempts to learn a French song. The Frenchman sings:

Vive Henri quatre
Vive ce roi vaillant!
Ce diable à quatre
Qui eut le triple talent
De boire, de battre
Et d'être un vert gallant

And, to the delight of his friends, the Russian mimicks the sounds:

Виварика́! Виф серувару!
Сидябляка́...
Кю... Кью-ю-ю... летриптала
Де бу де ба и детравагала (II, 586)[27]

Most striking, however, are the passages in which Tolstoj's aristocratic Russian protagonists converse in their macaronic Franco-Russian lingo.[28] The effect of this particular device cannot possibly be duplicated in translation because the very fact of speaking French had, in eighteenth- and nineteenth-century Russia, definite social, cultural, and even psychological connotations which the use of French did not quite carry in, say, England or Germany. French was not simply *the* language of the upper classes. Most members of the aristocracy employed

[26] The device was already used by Fonvizin in *The Minor* in the speeches of Mitrofanuška's German tutor, the illiterate former coachman Vral'man. Similarly, in Puškin's *The Captain's Daughter* an educated German officer in the Russian service is shown mispronouncing Russian.

[27] Again, an earlier instance of the use of this device may be found in Puškin's *The Captain's Daughter* (Chapter I) where the Russian serf Savel'ič mimics a French tutor. In fact, there is much similarity between Puškin's novel and *Don Quixote*. In both works we see two travellers, one of whom is a naïve nobleman (Puškin's is almost a child, while Cervantes' is child-like) and the other a peasant endowed with much common sense and *therefore* comic — both because his lower-class informality stands in sharp contrast to the dignified pose of his companion, and as the sober person who exposes the follies of his master. There is a definite resemblance between Savel'ič and Sancho Panza, as there is, incidentally, between the hidalgo's companion and two Gogolian heroes, the servant Osip in *The Inspector General* and the coachman Selifan in *The Dead Souls*. The kinship of the Spanish knight and Xlestakov in the Russian play and Čičikov in the novel is less pronounced.

[28] We are not discussing here Tolstoj's inconsistency in having the French characters in *War and Peace*, including Napoleon, use *both* French and Russian in their dialogue as this was not, obviously, intended to produce a comic effect. Nor shall we consider cases where Tolstoj's characters resort to foreign languages for purely utilitarian reasons, e.g., in order not to be understood by servants, etc. A more subtle instance of the practice is encountered in *Anna Karenina* (Part II, Chapter XXII) where Vronskij and Anna converse in French, "thus avoiding both the Russian *vy*, which was to them impossibly cold, and the dangerous [i.e., too obviously intimate] *ty*".

both languages in everyday life; the problem was the *quality* of the mélange, particularly of its French ingredient, a fact of which Tolstoj (as we learn from his autobiographical trilogy) was painfully aware at a very early age. Indeed, to young Tolstoj elegant French lexicon and, even more, impeccable pronunciation, were the chief criteria separating *les gens comme il faut* from those unworthy of this designation. In Tolstoj's day, all — or almost all — Russian noblemen spoke some French, a few spoke it quite well, but only a chosen élite could live up to Tolstoj's high requirements. Also, even if a Russian spoke French fluently, he could normally never forget that it was a foreign language. Hence the opportunity of using this bilingualism for comic effect.[29]

An effective example is provided by the opening lines of *War and Peace* where we meet *gens* who are definitely *comme il faut*:

— *Eh bien, mon prince. Gênes et Lucques ne sont plus que des apanages, des* поместья *de la famille Buonaparte. Non, je vous préviens, que si vous ne me dites pas que nous avons la guerre, si vous vous permettez encore de pallier toutes les infamies, toutes les atrocités de cet Antichrist* (*ma parole, j'y crois*) — *je ne vous connais plus, vous n'êtes plus mon ami, vous n'êtes plus* мой верный раб, *comme vous dites.* Ну, здравствуйте, здравствуйте. *Je vois que je vous fais peur,* садитесь и рассказывайте. (I, 5)
— *Je suis votre* верный раб, *et à vous seule je puis l'avouer.* Мои дети — *ce sont les entraves de mon existence.* Это мой крест. Я так себе объясняю. *Que voulez-vous*? (I, 10)

Occasionally, this kind of language is used ironically by the narrator himself, as for example, in Tolstoj's reference to a letter from Emperor Alexander to Napoleon, in which the French ruler is given the choice of "*verser* или не *verser le sang de ses peuples*" (II, 8).

The desire to speak "elegant" French often reaches absurdity. The Russian

[29] The tradition of "macaronic" comic verse is quite old in Russia. One of the classics is

Adieu, adieu, я удаляюсь
Loin de vous я буду жить
Mais cependant я постараюсь
Un souvenir de vous хранить.

There were even longer works in this genre, such as I. Miatlev's burlesque *Sensacii i zamečanija gospoži Kurdjukovoj* (1840). Although similar circumstances gave rise to similar literary phenomena in medieval Europe, where an individual's education (and consequently, his social standing) was frequently measured by his proficiency in Latin (thus, for example, there was a rich "macaronic" literature in one Slavic country, Poland, and echoes of the practice are much in evidence even in nineteenth-century Polish literature, particularly in works dealing with earlier periods, such as Adam Mickiewicz's *Pan Tadeusz* and Henryk Sienkiewicz's historical novels) the only recent European parallel that comes readily to one's mind is rather unexpected: Yiddish. Until the twentieth century, almost all speakers of the Yiddish vernacular also aspired to a knowledge of the sacred tongue, Hebrew. Many speakers of Yiddish were fond of quoting the Bible in the original Hebrew, but very few knew the language well, and comic distortions of the Biblical quotations were common. This fact was used with brilliant effectiveness by the great Yiddish humorist Sholom Aleichem.

nobleman Hippolyte, for example, believes that his reputation for speaking impeccable French would be enhanced by ... speaking Russian ungrammatically and with a foreign accent:

And Prince Hippolyte began to speak in Russian with the pronunciation of a Frenchman who had spent a year or so in Russia. ...
— В *Moscou* есть одна барыня, *une dame*. И она очень скупа. Ей нужно было иметь два *valets de pied* за карета. И очень большой ростом. Это было её вкусу. И она имела *une femme de chambre* ещё большой росту. Она сказала ... (I, 27)[30]

Another character in the novel, Šinšin, resorts to a diametrically opposite technique: he combines the most refined French with the most authentic "folksy" Russian. The result is no less amusing:

Ну как же, батюшка, *mon très honorable* Альфонс Карлыч, *vous comptez vous faire des rentes sur l'état*, с роты доходец получить хотите? (I, 71)

Sometimes Tolstoj resorts to outright mistranslation in order to heighten the comic effect of the statement. Thus, the French *légion d'honneur* is rendered as когорта чести instead of the correct почётный легион. Когорта (cohort) makes the institution appear ridiculous (II, 13). An entertaining scene is built around an encounter between Napoleon and a roguish Russian soldier who pretends not to recognize the French ruler; the Russian's shrewdly evasive answers to Napoleon's questions are mistranslated, and as a result the Emperor hears what he wants to hear (II, 132-33).

There is derisive laughter in Tolstoj's scenes which reflect Napoleon's vision of Russia as a primitive Asiatic land inhabited by exotic, semi-savage peasants. Thus, the cunning Russian soldier who had outwitted Napoleon himself was magnanimously set free by the French leader "*comme un oiseau qu'on rend aux champs qui l'ont vu naître*" (II, 133-34). Elsewhere, Napoleon, portrayed throughout the novel as a man with a theatrical awareness of the grandeur of his historical mission, is shown reflecting about

[30] Šklovskij (*op. cit.*, p. 212) notes that at least two characters in *War and Peace* — Count Miloradovič and an officer at the Russian headquarters — speak French poorly: the latter cannot decide whether to say *mon Monsieur Prince* or *Monsieur mon Prince*. Domestics were, of course, also eager to demonstrate their cultural superiority over primitive peasants, but, speaking no French at all, were forced to remain within the confines of Russian. The most common — and comic — method of such lexical ostentatiousness was the inappropriate and exaggerated usage of genteelisms. Thus, in a scene describing Pierre at his father's deathbed, — На другой бочок перевернуться хотят, — прошептал слуга и поднялся, чтобы переворотить лицом к стене тяжелое тело графа. (I, 99) Again, a parallel may be found in *The Captain's Daughter*, where Savel'ič muses "накушался бы чаю, почивал бы себе до утра" (Chapter I). Numerous examples of the device may also be found in various works of Dostoevskij, particularly the earlier ones, in countless stories of Leskov, and, among Soviet writers, in Babel' and, above all, Zoščenko.

Moscou, la capitale asiatique de ce grand empire, la ville sacrée des peuples d'Alexandre, Moscou avec ses innombrable églises en forme de pagodes chinoises! (II, 131)

In fact, in order to focus the Russian reader's attention on the Frenchman's preconceived silly notions of "savage" Russia, Tolstoj provides *French* translations of common *Russian* terms:

Увидав на той стороне казаков (*les Cosaques*) и расстилавшиеся степи (*les Steppes*) в середине которых была *Moscou la ville sainte* ... (II, 10)

Having conditioned the reader by a series of scenes at the beginning of the novel to regard German military science with much suspicion, Tolstoj later describes another military plan in German, even though he points out that this particular plan was *not* written in German. Thus the usage of the German language is in itself expected to evoke sarcastic scepticism:

Die erste Colonne marschiert туда-то и туда-то, *die zweite Colonne marchiert* туда-то и туда-то. (II, 466)[31]

In years to come, some of Tolstoj's comic devices and situations were even to be imitated by other Russian writers.[32] Thus, Maksim Gor'kij (on the whole, a rather humorless writer) recalls in the second part of his autobiographical trilogy his amusement caused by combinations of seemingly irreconcilable words. Among the instances of the oxymoron he cites such expressions as "до смерти хочу есть", "ужасно смешно", "страшно весело".[33] All this is reminiscent of Tolstoj's acid comments about Julie's reaction to verses describing death: to her they were "sweet" and "delightful" (I, 653).

Tolstoj's ironic portrait of Anna Pavlovna citing almost verbatim the Imperial

[31] Even in Tolstoj's loving description of his German tutor in *Childhood*, the use of German evokes gentle laughter. In *Anna Karenina* (Part VI, Chapter XXII), German is used ironically:
"— *Wünscht man Dochots, so hat man auch Klopots,* — сказал Васенька Весловский, подтрунивая над немцем.
"— *J'adore l'allemand,* — обратился он опять с той же улыбкой к Анне."
(Note: *Dochots* is a corruption of Russian доход).
It appears that Tolstoj was always vexed by orderlines and planning, both of which he considered to be German national traits; hence the "comic" application of the language. Unlike Mark Twain in *The Awful German Language*, Tolstoj did not ridicule the language itself, but the *Weltanschauung* he came to associate with it.

[32] We are not concerned here with instances of comic application of "serious" elements from Tolstoj's works by writers in years to come, i.e., with parody. Thus, it is not impossible that Vladimir Majakovskij's bad little boy Vlas Progulkin who, instead of being in school, amused himself by reading signs on stores from left to right and then from right to left ("Прочел с начала буквы он / Выходит "Куафер Симон" / Прочел с конца знаток наук / "Номис", выходит, "рефаук"") was actually imitating Pierre Bezuxov's mystic calculations of the ominous numerical significance ("666") of letters in the words *L'empereur Napoleon* and *L'Russe Besuhof* (II, 79).

[33] M. Gorkij, *Izbrannye proizvedenija v trex tomax* (Moskva: Goslitizdat, 1951), III, 249.

government's official statements on foreign policy at an "informal" soirée (I, 7) may have influenced numerous feuilletons of Mixail Zoščenko in which ordinary Soviet town folk parrot *Pravda* editorials and Party resolutions while squeezed in trolleycars, lined up in queues in front of grocery stores or suffering in dentists' waiting rooms. Tolstoj's caustic remark that *in spite* of physicians' efforts Nataša's health began to show some improvement (II, 68-69) is echoed in Zoščenko's "he's been treating me for five years. And everything's fine, my illness isn't any worse".[34]

The speech mannerisms of a Russian colonel in *War and Peace* who always substituted the hard "э" for the soft "е" (I, 76) migrated to the pages of Il'f and Petrov's *Little Golden Calf*, where Ostap Bender acquires a typewriter with the key for "е" missing, and is therefore forced to use the "э".[35] In fact — who can say? — one has a strong suspicion that Il'f and Petrov's beautiful, stupid, and cruel Elločka the Cannibal of the same novel is but a successive stage in the social evolution (perhaps the Russian измельчание would be a happier term) of Tolstoj's beautiful, stupid, and cruel Ellen.

The fact that Tolstoj's epic decries so much that is evil, senseless, and hypocritical has too often obscured the many varieties of laughter that may be found in it. Laughter is most often provoked by an unduly rapid shift of moods of behavior — whether it be Julie's instantaneous metamorphosis from a melancholic maiden into a very practical bride, or Bourienne's impartial readiness either to cry or to smile, or Napoleon's transformation from a haughty Emperor into a cowardly deserter, or the frequent transitions from the use of one language to another — reversals reminiscent of Hamlet's conversation with Polonius.[36] The countless instances of human folly related in the novel could not but evoke Tolstoj's indignation; it was equally inevitable that they should have produced laughter.

[34] Mikhail Zoshchenko, *Nervous People and Other Stories*, edited with an introduction by Hugh McLean (New York: Pantheon, 1963), p. 144.

[35] I, Il'f i E. Petrov, *Sobranie sočinenij v pjati tomax* (Moskva, 1961), II, 173.
Tolstoj's colonel: "Затэм, что император это знаэт. Он в манифэстэ сказал что нэ можэт смотрэть..." Bender's typewriter: "Отпуститэ податэлю сэго курьэру тов. Паниковскому для Чэрноморского отдэлэния на 150 (сто пятьдэсят) рублэй канцпринадлэжностэй..."

[36] Hamlet: Do you see yonder a cloud that's almost in shape of a camel?
Polonius: By the mass, and 'tis like a camel, indeed.
Hamlet: Methinks it is like a weasel.
Polonius: It is backed like a weasel.
Hamlet: Or like a whale?
Polonius: Very like a whale. (*Hamlet*, Act III, Scene 2)

TATIANA'S DREAM IN PUŠKIN'S *EVGENIJ ONEGIN*

VALDA NESAULE

Indiana University

Dreams play an important part in psychoanalysis; dreams are also used as literary devices. The purpose of this paper is to analyze Tatiana's dream — its content, its use to foreshadow coming events, and its symbolism.

Because the Russian language does not have a corresponding noun for the verb "to sleep" (*spat'*), the same word (*son*) is used both for "the sleep" and for "the dream". Besides these two literary meanings Puškin has given the word *son* a new meaning by using the word figuratively. Dmitry Čiževsky in the "Introduction" to *Evgenij Onegin* writes:

Too little attention has been given to "normal" Russian words used in the literary tradition without any special coloring — to which Puškin gave new meanings. These new meanings usually have lyric overtones, but they are sometimes bound up with a new *Weltanschauung*, a new outlook on life and poetry. The word "obman" for example ... The word "son" is another case. Puškin sometimes uses it in its ordinary sense ("sleep, dream"), but very frequently it comes to denote the "nocturnal side of the soul", spiritual depths and uncommon experiences drawing one outside the range of everyday life: "son voobražen'ja, son ljubvi, nadeždy temnyj son" (and even "bred"; cf. I, 58).[1]

According to *Slovar' jazyka Puškina* the word *son* is used in Puškin's works 379 times[2] and the word *snovidenie* 24 times.[3] In *Evgenij Onegin* the word *son* has various shades of meaning; the word is used twice in the stanza which describes Evgenij Onegin's undisturbed sleep before the duel (VI:xxiv), and it occurs three times in the two stanzas which show Onegin's dream-like condition after he has met Tatiana again (VIII: xx-xxi).

While dreams have often been seen as originating in the supernatural, nevertheless from ancient times on there has existed the theory that the dreamer

[1] Dmitry Čiževsky, "Introduction", Alexander Sergeevich Pushkin, *Evgenij Onegin* (Cambridge: Harvard University Press, 1953), pp. xxi-xxii.

[2] *Slovar' jazyka Puškina* (Moskva: Akademija Nauk SSSR, 1961), pp. 282-284.

[3] *Ibid.*, p. 238.

himself is the creator of his dreams. Thus Petronius Gaius, who was convinced that dreams are man-made, writes:

It is not the shrines of gods, nor the powers of the air, that send the dreams which mock the mind with flitting shadows; each man makes dreams for himself.[4]

Sigmund Freud interprets dreams as attempts to fulfill wishes;[5] Alfred Adler sees in dreams an expression of the dreamer's "style of life".[6]

If we agree that a man makes his own dreams, we may ask: What are the substantial elements of dreams? What is the "stuff" that "dreams are made on"? What is specifically the material used for the construction of Tatiana's dream?

The dream has a unified plot — Tatiana's journey to the dwelling place of her beloved, their meeting, and the sudden, tragic interruption of the short romantic episode. The dream consists of several parts, and each part has different elements. The atmosphere of the dream — snow, winds, darkness, and gloom — is similar to the atmosphere in Svetlana's dream in Žukovskij's ballad.[7]

At the beginning of her journey Tatiana encounters a hindrance — a dark torrent appears before her. There are two rods flung across the water, but they are slender and covered with ice (V:xi). Tatiana has to cross the dark, foaming torrent, but she is afraid to walk over the "*gibel'nyj mostok*". The sense of being doomed, of standing over an abyss overtakes Tatiana next night, the night after her name-day party when she is puzzled over Onegin's strange behavior:

... бездна / под ней чернеет и шумут (VI: iii)
... перед нею ... шумит ... поток (V: xi)

Earlier, in the third chapter, the author had predicted that Tatiana is destined to a tragic fate.

"Pogibneš, milaja", he says (III:xv). "Pogibnu", echoes Tatiana (VI:iii).

In her dream Tatiana sees that the small bridge across the torrent is shaky and perilous; nevertheless, she has to cross it.

As Tatiana stands helpless before the dark stream, sudden help appears; it is a shaggy bear who stretches out his paw to lead Tatiana across the icy bridge. The bear may be interpreted in several ways and may come from various sec-

[4] *Petronius*, The Loeb Classical Library (London: G. P. Putnam's Sons, 1919), p. 361.
[5] Sigmund Freud, *The Interpretation of Dreams* (New York: The Modern Library, 1950), p. 361.
[6] Alfred Adler, *What Life Should Mean to You* (New York City: Blue Ribbon Books, 1931), pp. 99-104.
[7] V. A. Žukovskij, "Svetlana", *Sočinenija*, I (Sanktpeterburg: Izdatel'stvo Glazunova, 1878), 213.

tions of Tatiana's mind. K. P. Laxotskij thinks that the bear comes from Russian fairy tales which Tatiana had heard:

Читая сон Татьяны, мы слышим дыханье русской народной сказки, начиная с типичного сказочного персонажа — медведя.[8]

A. Cejtlin writes about many painful experiences, many problems which Tatiana has to face sitting in front of Onegin's collection of books and standing on display in Moscow at the "brides' bazar".

Ей предстоит пройти через множество испытаний. У книжных шкафов Онегина, на Московской "ярмарке невест", в блестящем петербургском салоне образ Татьяны будет непрерывно развиваться, обогащаясь новым все более глубоким и жизненным содержанием.[9]

The friendly bear helps Tatiana to cross the dark current, but he does not let her continue her journey alone — he follows her. The next section of the dream is a typical "dream of being chased". J. A. Hadfield sees the source of these dreams in terror experienced in childhood.[10]

The bear in Tatiana's dream may be the symbol of the "*tolstyj general*" (VII: liv), the man who helps her to cross the abyss of despair and who carries her gently.

"Zdes' moj kum", says the bear (V: xv); Onegin is "rodnja i drug" of Tatiana's husband (VIII:xviii).

In both cases there is a relationship between the two, and in the dream as in real life Tatiana is placed on a "threshold" — she has a free choice between the two personages.

The fifth chapter of *Evgenij Onegin* has two main events: the dream of Tatiana and the description of her name-day party. Dmitry Čiževsky has expressed the view that there is a connection between the two subjects, and that the dream is a foreshadowing of some events which occur the next day at the party; in his "Commentary" to *Evgenij Onegin* he writes:

This duality of theme should have led interpreters of the novel to suspect that there was some connection between the two subjects. Yet until recent years comparatively few critics have observed this fact (M. Samarin, Gershenzon). The meaning of the second part is revealed by the symbolism of the dream.[11]

Both gatherings are noisy. There is a similarity in the description of the feast of the monsters and the party; the parallelism is specially obvious in the lines:

[8] K. P. Laxotskij, *Evgenij Onegin* (Leningrad, 1962), p. 42.
[9] A. Cejtlin, "Evgenij Onegin i russkaja literatura", *Puškin – rodonačal'nik novoj russkoj literatury*, ed. D. D. Blagoj and V. Ja. Kirpotin (Moscow and Leningrad, 1941), p. 341.
[10] J. A. Hadfield, *Dreams and Nightmares* (London, 1954), p. 168.
[11] Dmitry Čiževsky, "Commentary", *Evgenij Onegin*, p. 256.

Лай, хохот, пенье, свист и хлоп,
Людская молвь и конской топ! (V: xvii)

Лай мосек, чмоканье девиц,
Шум, хохот, давка у порога. (V: xxv)

Tatiana listens to the senseless noise in the little hut and watches the monsters through a peep-hole. She is frightened. Then she recognizes the host of the feast; it is the dreaded and yet beloved man — Onegin. Gukovskij interprets Tatiana's dream as a typical dream told by the bride on the morning of her wedding day in Russian folk songs:

Мотивы этого сна осмысляются ... как "пророческие", как предвидение будущей судьбы невесты, и избушка означает дом новой семьи, а чудовища — новую семью ... Таким образом можно думать, что Пушкин в сне Татьяны объединил разнообразный фольклорный материал композиционной формулой ритуального сна невесты.[12]

The recognition of Onegin as the host of the wild feast has several levels of significance. The company which he is keeping casts a shadow over him, and perhaps at this point Tatiana starts to question Onegin's integrity and perhaps here starts the feeling of a slight disenchantment which is later expressed in the question:

"*Už ne parodija li on?*" (xxiv: VII)

There is, however, the possibility of a different evaluation. K. P. Laxotskij writes that Tatiana did see Onegin superior to all other men and that this is reflected in the fact that Onegin is the one who commands and all monsters have to obey.[13]

One way or the other, whether Tatiana still places the image of Onegin above all other men, or whether she becomes uncertain, does not matter; he is and remains the man of her destiny, the one whom she loves and will continue to love (VIII: xlvii).

The next part of the dream is a typical wish-fulfilment dream. Tatiana is not afraid any more; she herself opens the door. Suddenly there is a blow of the wind, the lights go out, and Onegin claims his victim.

"Моё", cries Onegin (V: xx).

It is significant that here the impersonal neuter pronoun is used instead of the feminine "moja".

Now comes the climax of the dream: the journey is ended, the wish is fulfilled, the mighty hero takes the trembling girl and places her gently on a bench. They are alone. This short moment of happiness is reflected next day when

[12] G. A. Gukovskij, *Puškin i problemy realističeskogo stilja* (Moskva: Gosudarstvennoe izdatel'stvo, 1957), pp. 216-217.
[13] K. P. Laxotskij, *op. cit.*, p. 42.

Onegin bows silently to Tatiana and his look expresses deep tenderness:

Он молча поклонился ей;
И как-то взор его очей
Был чудно нежен. (V: xxxiv)

This short dream of happiness is, in a way, fate's gift to Tatiana. The happiness which she never experienced in real life is achieved now in a dream. This dream episode is perhaps created by Tatiana as a device of the defense mechanism, whereby her ego is protected against the shame and disgrace which her rejected love had brought upon her. This "creation of the mind" becomes a "wild reality".[14]

The short romantic episode is interrupted by unbidden guests, Ol'ga and Lenskij. A quarrel starts and Lenskij falls, hit by Onegin's knife. Tatiana awakens in terror.

The last episode of the dream is a typical foreshadowing dream which foretells the tragic death of Lenskij. The possibility of preventing the duel is expressed in VI: xviii:

Когда бы ведала Татьяна,
Когда бы знать она могла,
Что завтра Ленский и Евгений
Заспорят о могильной сени;
Ах, может быть ее любовь
Друзей соединила б вновь!

Tatiana believed that dreams foretell the future:

Татьяна верила преданьям
Простонародной старины,
И снам и карточным гаданьям. (V: v)

The horror of Onegin's quarrel with Lenskij causes Tatiana to wake up. She searches in her dream book for the hidden meaning of the alarming dream, but the book can not give her a clear answer:

Ее сомнений
Мартын Задека не решит,
Но сон зловещий ей сулит
Печальных много приключений. (V: xxiv)

The last part of the dream becomes again a device of Tatiana's defense mechanism — the cause of the duel is changed, and the duel's tragic outcome is made the factor which prevented Onegin's and Tatiana's eventual marriage. Although before the dream Onegin had told Tatiana that he did not love her, Tatiana makes him in her dream a lover who is separated from her by tragic events.

[14] Lord Byron, "The Dream", *The Selected Poetry of Lord Byron* (New York: The Modern Library, 1951), p. 326.

Years later Onegin repeats the same version; he writes in his letter to Tatiana that Lenskij's death was one of the causes of their separation:

Еще одно нас разлучило.
Несчастной жертвой Ленский пал. (VIII: xxxii)

A dreamer is aware of the problems and thoughts that occupy him in waking life. "The wounds of the unhappy endure throughout the night", writes Petronius.[15] Sometimes in a dream "the wounds of the unhappy" are more painful than they are during the day. The dreamer attempts to create a solution to his problem, and, since so much of thinking during the sleep is done in visual images, the solution to the problem is expressed in images.

Tatiana's wounds are deep. There is not only the problem of her unanswered love, but there is also a painful feeling of shame similar to Kitty's "agony of shame" in Tolstoj's *Anna Karenina.*[16] Tatiana is ashamed that she has written the letter confessing her love to Onegin. In her dream Tatiana attempts to solve her painful problem by making it appear that Onegin loves her, that their happiness is near, and that Lenskij and Ol'ga are the destroyers of her union with Onegin.

"A sčast'e bylo tak vozmožno, tak blizko", says Tatiana to Onegin in the eighth chapter (VIII: xlvii).

Tatiana does not understand the symbolical meaning of her dream and is unable to bring the structural elements of it into a meaningful relationship. But dreams are meaningful even when they are not understood.

"Dreams are parables, symbolic stories, carrying a deeper meaning", writes J. A. Hadfield, "and they have an effect on the mind even when they are not consciously understood or interpreted".[17]

Tatiana's dream is meaningful as a literary device. G. Gukovskij writes that the use of almost eleven stanzas for the description of the dream would be a mistake if it were not for the fact that the dream reveals the inner life of Tatiana.

То обстоятельство, что Пушкин решился пойти на риск задержки хода романа, показывает, насколько важен был для его замысла, для его идеи сон Татьяны. ... Сон Татьяны заменяет у Пушкина подробный анализ ее психологического мира воплощая его в образах ... это ключ к пониманию ее души ...[18]

Tatiana's dream may be interpreted in several ways; it has "a wide realm of wild reality"; it is "the key to Tatiana's soul".

[15] *Petronius, op. cit.*, p. 362.

[16] Leo Tolstoy, *Anna Karenina*, Part I, Ch. 22: "Kitty looked into his face, which was so close to her own, and long afterwards — for several years after — that look full of love, to which he made no response, cut her to the heart with an agony of shame."

[17] J. A. Hadfield, *op. cit.*, p. 102.

[18] G. A. Gukovskij, *op. cit.*, p. 214.

THE ANALOGICAL SPREAD OF RUSSIAN /o/ AND /e/

WILLIAM R. SCHMALSTIEG

The Pennsylvania State University

The analogical innovation which seems on the least secure footing is the one which Hockett calls *allophonic analogy.* According to him, "in allophonic analogy one allophone of a phoneme comes to occur in positions previously open only to some other allophone of the same phoneme, and a new phonemic contrast is thus created".[1] Hockett gives the example of Caribbean Spanish, in which the allophone [h] of the phoneme /s/ which was originally limited to word final position and position before a consonant has spread beyond its original limits to occupy the position before a vowel as well. The purpose of this article is not to discuss Caribbean Spanish, but I should like to point out that this example of Hockett's is not on very solid ground. My colleague, Prof. Ricardo A. Narváez of the University of Minnesota, maintains that the phonetic [h] of Caribbean Spanish is still an allophone of /s/. The [h] occurs only at the end of the syllable whereas the [s] occurs in syllable initial position.[2]

As his second example Hockett cites the case of English /θ/ which split into two phonemes, viz. /θ/ and /ð/ by way of the restressing of atonic words such as *this, that* and *the.*[3] On the other hand, in terms of distinctive features no new contrast has been added. The distinction between voiced and unvoiced consonants had existed all along and the distinction between the dental spirant and the other consonants was not new to the language either. Thus from the standpoint of the system of the language it meant merely filling a hole in the pattern.

The phenomenon is similar to that of the development of /f/ in Russian. Originally [f] was merely an allophone of /v/ which was unvoiced in certain positions. Since, however, voicing vs. the lack thereof was distinctive in certain

[1] Charles F. Hockett, *A Course in Modern Linguistics* (New York, 1958), p. 434. This is in marked contrast to the view of Roman Jakobson, who says, "*il n'existe pas, dans le cadre d'un système appartenant à un seul style de langage, de phonologisations qui ne soient pas compensées par des déphonologisations*" ("Principes de phonologie historique" printed in Cantineau's translation of Trubetzkoy's *Grundzüge der Phonologie* [Paris, 1949], p. 332).

[2] Prof. Narváez' article, "Algunos comentarios sobre la pronunciación del Castellano en Puerto Rico", appeared in *Cultura,* No. 114/115 (Jan.-Sept., 1963), pp. 213-218. This journal is published in Tunja, Boyacá, Colombia, South America.

[3] Hockett, *op. cit.*, pp. 430, 434.

other positions and the labio-dental spirant already existed as a phoneme, the way was open for /f/ to appear as a full-fledged phoneme. It was only a new combination of distinctive features which already existed in the language.

Let us consider another example of *allophonic analogy* taken from the Slavic field. The traditional handbooks of the Slavic languages state that in Russian an *-e-* passed to *-o-* following a palatalized and preceding an unpalatalized consonant.[4] A better analysis of the situation is presented by Horace G. Lunt who says, "In most of northern East Slavic [e] (< *e*, *ь*) existed only between soft consonants; [ö] (< *e*, *ь*) followed a soft consonant in the absence of a succeeding soft; while [o] was found after a hard consonant, so that the single mid vowel phoneme had positional variants C'eC' ~ C'o(C) ~ Co(C). It was not for some time that morphological analogy and/or the change of *ě* > *e* disturbed this distribution, so that either C'oC' or C'eC became possible and the unrounded *e* was phonemically opposed to rounded *o*."[5] R. I. Avanesov recognizes the fact that *o* and *e* were members of the same phoneme, but says that in such forms as *na berëze, na klëne, nesëte* the /o/ (represented orthographically by *-ë-*) became possible by analogy with forms such as *berëza, klën, nesëm*.[6] Now if it were not for the play of analogy it could be only the passage of */ě/ to /e/ which called forth the contrast between /e/ and /o/. But this contrast could have arisen only in the group C'eC vs. C'oC or in word final position. In the group C'eC' and CoC there would have been no contrast; between unpalatalized consonants only /o/ would have been possible and between palatalized consonants only /e/ would have been possible. In other words the contrast between /e/ and /o/ should have been neutralized in these positions. On the other hand, it is clear that the contrast was not neutralized, but actually spread beyond its original borders to forms such as *na berëze*, *na klëne* and *nesëte*.

It should be pointed out here that a contrast which is possible in one position is not *necessarily* free to spread to other positions. Thus the contrast between voiced and voiceless consonants is neutralized in word final position in Russian and German. Although in the declension there would seem to be every morphological pressure to do so, analogy has not reintroduced this contrast in word final position, cf. Russ. nom. sing. /got/ 'year', gen. sing. /goda/ or German nom. sing. /tot/ 'death', gen. sing. /todəs/.

How is it possible then that the contrast between /e/ and /o/ has widened its distribution so that both C'oC' and CoC have become possible? I think that the matter can be explained if we consider the distinctive features of Russian

[4] André Vaillant, *Grammaire comparée des langues slaves*, I (Paris, 1950), 108; P. Ja. Černyx, *Istoričeskaja grammatika russkogo jazyka* (Moscow, 1952), p. 115.

[5] "On the Origins of Phonemic Palatalization in Slavic", *For Roman Jakobson* (The Hague, 1956), p. 312.

[6] *Očerki russkoj dialektologii* (Moscow, 1949), p. 43.

from the articulatory point of view. In the high vowels the feature of lip rounding was possible between palatalized consonants in Russian, cf. *ljubit'* 'to love', *ljudi* 'people'. The phoneme /u/ was not limited in distribution in the same manner as the phoneme /o/; /o/ originally could not occur between palatalized consonants, whereas /u/ could. The extension of /o/ to the position between palatalized consonants requires no new arrangement of the distinctive features. There is no magic in this. The articulatory organs were accustomed to producing /o/ and they were accustomed to producing rounded vowels between palatalized consonants. The new phonemic arrangement (i.e. the appearance of /o/ between palatalized consonants) makes no new demands on the articulatory organs. No distinctive features begin to appear in positions where they had not occurred before.

The passage of */ě/ to /e/ and the phonemic split of */e/ into /o/ and /e/ led to a new distribution of phonemes such that all of the phonemes could occur after unpalatalized consonants except /e/. Thus /Ci, Ca, Co, Cu, C'i, C'e, C'a, C'o, C'u/ were possible but not */Ce/. This distributional lack has been remedied by the introduction of foreign words such as /ten'is/ 'tennis'. Again this balancing of the system did not require the introduction of any new distinctive features. The unrounded phoneme /i/ (phonetically [y]) was possible after unpalatalized consonants, the mid vowel /o/ was possible after unpalatalized consonants; therefore the unrounded mid vowel /e/ also became possible in this position. In terms of the phoneme the distribution is new; in terms of distinctive features the distribution is old.

The conclusion is that analogy in Russian has introduced new arrangements of phonemes, but it has not introduced a new arrangement of distinctive features.

THE POLISH NAPOLEONIC CULT FROM MICKIEWICZ TO ŻEROMSKI: A RAPID SURVEY

H. B. SEGEL

Columbia University

> But what is this outcry that rises to heaven,
> Whom are you greeting, O nations?
> It is a god, who is bringing us freedom,
> He is coming in all his majesty! ...
> He has greatness in his heart, sweetness
> in his face,
> Thunderbolts for oppressors are glowing
> in his hand!
> Give greeting, surprised earth,
> To the superhuman soul of Napoleon!
>
> Franciszek Wężyk (1785-1862) in
> *An Ode to Poles, Written in the Month of December*, 1806.

> Poland exists only in the imagination of those who want to use it as a pretext for spinning dreams.
>
> Napoleon

Mickiewicz left Russia in 1829 and travelled through Western Europe before finally settling down in Rome. It was here that news reached him of the outbreak of the ill-fated November Insurrection of 1830. The poet's reaction was not one of exultant enthusiasm. Perhaps he did feel he should participate in the struggle for Polish independence — it was an expected gesture — but his experiences and observations in Russia made him painfully aware of the folly of Polish attempts to liberate their country by force of arms. At some point, however, Mickiewicz made up his mind to go to Poland, but his itinerary was hardly direct. From Rome he went first to Paris, where he dallied a few months. Then came Dresden, and finally Poznań, which he reached in time to learn of the fall of Warsaw to the Russians. Mickiewicz lingered in Poznań almost eight months, obsessed by feelings of guilt over his non-participation in the Insurrection. In March 1832, he left the Poznań area for Dresden, where he identified himself closely with the mass of Polish émigrés heading westward in what since has become known in Polish as the *Wielka Emigracja* (The Great Emigration). Once the emigration had settled in Paris, Mickiewicz assumed the role of the *wieszcz* or poet-seer who is the spiritual leader of his people. It was a role to which his whole previous career had in fact inclined him. Its realization, however, needed the proper catalyst and that was provided in the form of the November Insurrection.

The three great works of Mickiewicz's most fertile years as a creative artist — *Dziady, część trzecia* (Forefathers' Eve, Part Three), *Księgi narodu i pielgrzymstwa polskiego* (The Books of the Polish Nation and Pilgrimage), and *Pan Tadeusz* — were all written in a relatively brief period of time — 1832-1834 — when, as *wieszcz*, Mickiewicz's power was at its height. In their spiritual cohesion these three works represent the three acts of the poet's personal drama as *wieszcz*. In the first, *Forefathers' Eve*, Mickiewicz made the necessary *apologia* before draping himself with the mantle of *wieszcz*. It was an act of self-expiation for his non-involvement in the great upheaval of November 1830, and it took the form of a hyperbolic dramatization of the investigation and subsequent trial of the Wilno Philomathians which led to Mickiewicz's four-year exile in Russia. By showing that he too had suffered for the national cause — despite the fact that he did not actively participate in the November Insurrection — the poet established (or attempted to establish) the validity of his claim to the recognition and acceptance of himself as *wieszcz* by the masses of his fellow émigrés. In the second act of the drama, *The Books of the Polish Nation and Pilgrimage*, Mickiewicz now for the first time fully in the role of *wieszcz* expounded two of the basic tenets of the Polish émigré faith: a) Poland's messianic destiny, and b) the concept of the emigration as a pilgrimage. In the second part of the work, "The Books of the Polish Pilgrimage", the poet equipped the émigrés with a guide in the form of biblical-style parables on the way they were to conduct themselves during their "pilgrimage".

The last act of the drama and the greatest, *Pan Tadeusz*, was a communal rite of magic by which all the rivalries and divisions of the emigration were to be made to vanish before the conjured image of a common heritage. For all its beauty the poem is, on one level of interpretation, a call for the émigré-pilgrims to forget the sharp differences that had sprung up in the emigration and to join together in common cause for the common good. Just as the feuding Horeszkos and Soplicas forget their differences in the face of the common enemy — the Russians who arrive to put an end to their foray — so too were the émigrés to present a common front to the enemies of despair, frustration, and antagonism which the Emigration engendered and which Mickiewicz's only real rival for the mantle of *wieszcz* — Juliusz Słowacki — portrayed so vividly in *Anhelli*, a work written as an answer to and refutation of the theses propounded by Mickiewicz in *The Books of the Polish Nation and Pilgrimage*.

One aspect of this common heritage evoked by Mickiewicz no less profoundly than others in *Pan Tadeusz* and which was to become along with messianism and the emigration-pilgrimage equation a main current in the Polish Romantic émigré ideology was the worship of the Emperor Napoleon and the Polish Napoleonic alliance. As we shall see later, it was this conscious elaboration of the Napoleonic cult that eventually led to the humiliation of Mickiewicz's resignation under pressure from the Chair of Slavic Literatures that he had held in the Collège de France between 1840 and 1844.

The Polish Napoleonic adventure had been initiated in 1797, two years after the Third Partition, when Polish Legions (*legiony*) were formed under the command of General Henryk Dąbrowski and incorporated into the French army in Lombardy. Although some voices inevitably were raised in opposition (Kościuszko's for example), the majority of Poles firmly believed that Napoleon offered them the only hope of liberating Poland from the oppression of the partitioning powers, and also (and no less meaningful to them) of enjoying once again the dignity of fighting as free men for a cause they believed in with all their hearts and souls. By joining Napoleon they were striking not only a blow for Polish freedom, but, of more universal significance, a blow for freedom of oppressed peoples everywhere. Not many years elapsed before they were to discover the bitter irony that the Emperor had destined them for quite contrary purposes. Thousands of Poles flocked enthusiastically to Napoleon's banners and fought and died under these banners from one end of Europe to the other and beyond — in distant Santo Domingo.

After the victory at Jena in 1806, the Emperor sent out a general call for the Poles to join him. A large force was organized and was entrusted to the command of the gallant soldier Prince Józef Poniatowski and General Dąbrowski. The success at Danzig, Preussisch-Eylau and Friedland culminated in the Peace of Tilsit. As a gesture of good will to the Poles, and as preparation for his further ambitions in the East, Napoleon ushered into existence in 1807 the short-lived Duchy of Warsaw (Księstwo warszawskie) made up of territories seized by Prussia in the second and third Partitions. And it was from this partial and lamentable fulfillment of the Polish dream of liberation that the forces were recruited that helped crush the popular resistance movement in Spain in 1808 and achieve the liberation of Galicia from Austrian rule the following year. Buoyed up by the victory in Galicia (which somehow could justify the cynical use to which they had been put in Spain) the Poles eagerly anticipated the titanic struggle with Russia that loomed on the horizon of the year 1812. The rest of the political history is too well known to recount here. Polish hopes crumbled in the defeat of the *Grande Armée*; the humiliation of Santo Domingo and Saragossa seemed to lose whatever meaning they might have had, and what was left was an almost unutterable despair and anguish that still could not bring itself to a total renunciation of the myth of Napoleon the Liberator, the Saviour.

This then was the raw historical material out of which the Polish Napoleonic cult was created. The Napoleonic adventure ended decisively at the Congress of Vienna in 1815; seven years later Mickiewicz's *Ballady i romanse* (Ballads and Romances) appeared heralding the advent of Polish romanticism. It was the Romantic movement that hastened the spread of the Napoleonic myth.[1] Throughout Europe the Romantics deified the Emperor; to them he was the embodiment of the spirit of rebellion, he was the Hero, the maker of history, the incarnation of the ideals and aspirations of their age. It was a worship that crossed national boundaries and from which few could remain aloof. On the Poles, who had placed their political fortunes in Napoleon's hands, it left an indelible stamp.

The Polish Napoleonic myth was intimately bound up with the figure of Mickiewicz, its most ardent and eloquent champion. Prior to his assumption of the role of *wieszcz*, there is no evidence that Napoleon had much impact on the poet's life. There were of course the stirring boyhood recollections of the passing of the Emperor's legions through his birthplace, the small village of Nowogródek in Polish Lithuania; there were the patriotic and virile poems of the legionnaires who had fought with Dąbrowski and the virtues of which Mickiewicz was to extol later in his lectures at the Collège de France. But before the Emigration, before the period of his *wieszczostwo* and specifically *Pan Tadeusz*, Napoleon intrudes nowhere in Mickiewicz's creative work.

The impetus to the elaboration of Mickiewicz's personal Napoleonic faith in the period 1832-34 when *Pan Tadeusz* was written came from two sources principally: a) the rapid growth of the Napoleonic legend amongst the Romantics after the Emperor's death in 1821, and b) of even greater meaning for Mickiewicz and the Polish Emigration, the official evocation and elaboration of the *légende napoléonienne* by the bourgeois monarchy of France presided over between 1830 and 1848 by Louis-Philippe (1773-1850). The historian of modern France, Albert Guerard, has characterized the Napoleonism of Louis-Philippe's reign as an outlet, together with the conquest of Algeria, for a glory-starved bourgeois population.[2] The first tangible symbol of this revival of the Napoleonic cult in the time of the bourgeois monarchy was the restoration in 1833 of the statue of Napoleon on the top of the Column in the Place Vêndome. The next was the completion of the Arc de Triomphe. The last, and the most impressive, was the "Return of the Ashes" (Le Retour des Cendres) in 1840, when the incongruously named frigate, *La Belle Poule*, brought back the sacred remnants

[1] On the Napoleonic myth, see Albert J. Guerard, *Reflections on the Napoleonic Legend* (London, 1924), and J. Lucas-Debreton, *Le Culte de Napoleon* 1815-1848 (Paris, 1960).

[2] *France, A Modern History* (Ann Arbor, 1959), pp. 292-293.

from St. Helena. The Great Emigration coincided then with the official Napoleonic policy fostered in the first ten years of its reign by the government of Louis-Philippe, and from the meeting of the two there emerged the richest fruit of Polish Napoleonism, Mickiewicz's *Pan Tadeusz*.

The Napoleonic campaigns in *Pan Tadeusz* form the epic backdrop against which the homely routine of life in the Polish-Lithuanian gentry village of Soplicowo is played out. But gradually the campaigns draw nearer to this remote corner of Lithuania. Eventually the two merge as the Emperor's legions enter the village for the great celebration and rejoicing that brings the poem to an end. The means by which Mickiewicz prepares the advent of the Napoleonic legions in Soplicowo is Jacek Soplica who in the form of Father Robak has devoted his life to the expiation of the sin of the murder he committed in a moment of impetuous youthful anger. Robak appears in the novel as an agent of Napoleon come to Lithuania to rouse the Polish gentry to support of the Emperor,

> that man, the god of war who surrounded by a cloud of regiments, armed with a thousand cannon, harnessing to his chariot golden eagles beside those of silver, was flying from the deserts of Libya to the lofty Alps, casting thunderbolt on thunderbolt, at the Pyramids, at Tabor, Marengo, Ulm, and Austerlitz. Victory and conquest ran before and after him. The glory of so many exploits, heavy with the names of heroes, went roaring from the Nile to the North, until at the shores of the Niemen it was beaten back as from crags by the Muscovite lines that defended Lithuania as with walls of iron against tidings terrible for Russia as the plague.[3]

In Books XI and XII of *Pan Tadeusz* Mickiewicz brings the Napoleonic theme into the foreground, that is, after all local problems have been resolved, particularly the contest of wills between the rival Horeszko and Soplica clans. The armies of Napoleon arrive to shatter the tranquility of Soplicowo. Some of the great Polish generals are there: Dąbrowski, Kniaziewicz, Pac, Małachowski. The atmosphere is exhilarating, expectant: great victories lie ahead and with them the defeat of Russia and the rebirth of Polish independence. To celebrate the occasion a magnificent banquet is laid out during which the patriotic Jewish inn-keeper Jankiel comes forth to entertain the assembly of gentry-folk and soldiers with a musical review of memorable Polish historical events beginning with the Constitution of May 3, 1791, and concluding with the Mazurek Dąbrowskiego of Wybicki, the Polish legionnaires' song that has since become the Polish national anthem. The banquet draws on into the evening, the sun sets — and *Pan Tadeusz* ends.

Mickiewicz's fervid Napoleonism, which harbored few doubts, dictated the

[3] Adam Mickiewicz, *Pan Tadeusz*, trans. George Rapall Noyes, Everyman's Library, No. 842 (London, 1949), p. 30.

casual treatment of history that we find in *Pan Tadeusz*. We know, for example, that Napoleon definitely decided on war with Russia in the spring of the year 1812, and that the campaign was in fact begun in late June. In *Pan Tadeusz*, however, the decision is moved up to the late summer of 1811 so that the dying Robak can learn of the successful completion of his mission. The *Grande Armée* passed through Lithuania in the summer of 1812, but again Mickiewicz pushed the time of action back to the spring of 1812 in order to exploit the "rebirth" symbolism of the spring season. Furthermore, Mickiewicz presents the Polish-Lithuanian gentry as solidly backing the campaign against Russia. This was hardly the case. To seduce the gentry away from support of Napoleon, the Russian tsar (Alexander I) offered concessions in the form of increased autonomy to Lithuania under Polish rule. His efforts were not wholly unsuccessful and a number of the gentry looked with disfavor on Polish cooperation with the French. Yet in *Pan Tadeusz* there is only one dissenting voice: that of the old village sage Maciek of the Dobrzyński clan. When General Dąbrowski turns to him and asks:

What have you to say, comrade? Can it be that you are not glad at our coming? Why are you silent and dumb? How can your heart help leaping when you see the gold and silver eagles, and when the trumpeters trumpet Kościuszko's reveille close to your ear? Maciek, I thought that you were more of a fighting man: if you do not seize your sabre and mount your horse, at least you will gaily drink with your colleagues to the health of Napoleon and the hopes of Poland.[4]

Maciek is less than enthusiastic. He answers Dąbrowski:

But, sir, two eagles never nest together! Lords' favour, hetman, rides with a piebald steed! The Emperor a great hero! On that subject we could expend much talk! I remember that my friends the Puławskis used to say, as they gazed on Dumouriez, that Poland needed a Polish hero, no Frenchman or Italian either, but a Piast, a Jan or a Józef, or a Maciek — that's all! The army! They say it is Polish! But these fusilers, sappers, grenadiers, and cannoneers! You hear, in that crowd, more German than native titles! Who can understand them! And then you certainly must have with you Turks or Tartars or Schismatics, men without God or without faith: I have seen it myself; they are assaulting the peasant women in the villages, plundering the passers-by, pillaging the churches! The Emperor is bound for Moscow! That is a long road if he has set out without the blessing of God.[5]

The reasons behind Maciek's dissatisfaction are understandable. The army going into the campaign against Russia is not native and therefore cannot be dedicated chiefly to Polish interests. Then, he is afraid that it is "godless" (Maciek must have heard something about Napoleon's difficulties with the Papacy). These are not serious objections but they spring from the mouth of a

4 *Ibid.*, p. 313.
5 *Ibid.*, pp. 313-314.

respected village elder who is meant to be taken seriously. The skepticism of Maciek may be an externalization of similar doubts on the part of Mickiewicz himself. But they are presented as unique in *Pan Tadeusz*; there is no other dissenting voice to be heard except that of Maciek, and even his is lost in the sea of general acclaim that accompanies the arrival of the Napoleonic forces in Soplicowo. The over-all mood is one of joy, and it is precisely for this reason that *Pan Tadeusz* ends before it would have become necessary for Mickiewicz to present any of the grim details of the debacle that befell the Napoleonic campaign against Russia. As Wiktor Weintraub has so accurately described it:

> There is not the least premonition to the effect that the Napoleonic campaign will end in a debacle. We enter triumphantly into an era of full happiness crowned with political liberty and social justice, and this era is introduced as though it will last forever. We are here under the sway of a daydream. Verweile doch, du bist so schön![6]

Thus, the conclusion of *Pan Tadeusz* is entirely in keeping with the spirit of the work as a whole and Mickiewicz's own Napoleonic hero-worship.

This worship reached a point of culmination in the last two years of Mickiewicz's professorship of Slavic Literatures at the Collège de France: 1842-44. By now the poet's literary masterpieces were behind him; he had become caught up in the Polish émigré mystic cultism that swirled about the enigmatic figure of the Lithuanian "priest" Andrzej Towiański. Mickiewicz fell completely under Towiański's sway, and in his last two years at the Collège de France his lectureship served no longer as a platform from which to dispense information and opinion about the cultures and literatures of the Slavic peoples, but as a pulpit from which to preach a Napoleonic worship shot through with mystic ecstasy and symbolism. It began approximately on Tuesday April 29, 1842, in the twenty-second lecture for the academic year with Mickiewicz's review of the poets of the Duchy of Warsaw Kajetan Koźmian, Franciszek Wężyk and, more importantly, the "soldier-poets" — former legionnaires — Cyprian Godebski, Wincenty Reklewski, and Antoni Gorecki, with whom chronologically the vast Polish Napoleonic literature truly begins, It ended dismally in the last lecture at the Collège de France on Tuesday May 28, 1844, when Mickiewicz passed out among his audience lithographs prepared by the young French artist, Tony Toulion, showing the symbolic portrait of Napoleon in a long mourning veil with his eyes raised upwards, a pained expression on his face and indicating with his hands a map of Europe spread out before him. Appropriately the painting on which the lithograph was based was the work of a member of the Towiański Circle, Walenty Wańkowicz, whom Mickiewicz had known in Wilno and later also in St. Petersburg. After distributing the lithographs, Mickiewicz called on

[6] *The Poetry of Adam Mickiewicz* ('s-Gravenhage, 1954), p. 246.

his audience to invoke the spirit of "their Hero, the Man of Destiny", who alone could lead them to the "appointed man". By the last, the poet doubtless meant Towiański himself, whom he saw as the continuer of Napoleon's spiritual mission on earth:

This portrait is a sign by which you shall know him. I entrust this portrait to your memory. For me it shall be testimony that I have fulfilled my responsibility. The day will come when you will be called to an accounting of how you fulfill your responsibility.[7]

A toast to Napoleon brought the session — and Mickiewicz's lectures — to an end.

The deification of Napoleon that put its stamp on the early years of the bourgeois monarchy of Louis-Philippe became a source of embarrassment to the "citizen king" in the half decade of growing dissatisfaction and insurrection preceding the revolutions that toppled the monarchy and ushered in the Second Republic (1848-1852). In these circumstances, the use of the Chair of Slavic Literatures at the Collège de France for the continued propagation of a mystical neo-Napoleonism in the mid 1840's was unthinkable and Mickiewicz was gently eased out of the position (the official excuse was a "leave of absence"). But by this time, Mickiewicz's worship of the Emperor, which went so far as to hail Napoleon as a Messiah, had alienated not only much of his public, but even his closest friends among the faculty at the Collège de France. J. Lucas-Debreton points this out in his excellent study *Le Culte de Napoleon 1815-1848*:

Professeur au Collège de France, il proclamera Napoléon le saint de l'époque, le grand réalisateur dont le culte doit réunir tous les hommes d'action; et resté catholique, il se séparera de Michelet, de Quinet, républicains anticléricaux, par cette vénération pour le nouveau Messie qui attend "son successeur spirituel".[8]

The Polish Napoleonic cult was preserved most zealously now in the ranks of the émigrés who constituted the membership of the Towiański circle in which Mickiewicz assumed an increasingly important role following his departure from the Collège de France. Napoleon worship was central in Towiański's mystic system and represented a cornerstone of the Towianist faith. For Mickiewicz, it was an ideal to which the poet remained faithful until his death in 1855.

The eruption and subsequent defeat of the January Insurrection of 1863 in Russian Poland sounded the death knell of Polish political romanticism and with it the cult of Napoleon. The Romantics' inflammatory writings and utterances were held responsible to a great extent for the ill-conceived, awkwardly-executed revolts of 1830 and 1863 that had little or no hope of success and only brought

[7] Adam Mickiewicz, *Dzieła*, XI, ed. Leon Płoszewski (Warsaw, 1955), pp. 509-510.
[8] Lucas-Debreton, *op. cit.*, p. 415.

an increase in Polish misery in their wake. The ideology of positivism that supplanted romanticism after 1863 had little room and even less sympathy for the political beliefs fostered by the Romantics and vigorously combatted them through journalism and fiction. Revolution was rejected as a means of achieving Polish independence in favor of the evolutionary process that called first for an improvement of all aspects of Polish life within the context of partition. The emphasis was on schools, not guns; on social and economic development, not mystic dreams of messianic deliverance and Napoleonic conquest. The leading positivists — Świętochowski, Prus, Orzeszkowa, Asnyk — struck out at romanticism in its diverse forms in one work after the other. Attention shifted away from the émigrés whose chances had come and gone to internal Polish developments and concerns. Whatever smacked of a lingering romanticism had to be uprooted if Polish society after 1863 was to move forward to a realistic appraisal of itself and its relations with the partitioning powers. One vestige of romanticism which the Positivists viewed as particularly pernicious because of its hold on the Polish consciousness was Napoleonism, and they did their best to expose it as the tragic folly they believed it was. The case is vividly stated in the outstanding novel of Polish positivism, Bolesław Prus's *Lalka* (The Doll), a gigantic panorama of post-1863 Varsovian life.

As the weapon for his deromanticization of Napoleonism in the novel Prus chose the journal kept by the shop-keeper Ignacy Rzecki. This novel within a novel ("Pamiętnik Starego Subiekta", in the original Polish) serves two highly important purposes. One is structural, the other ideological. Structurally, the memoirs are a *Vorgeschichte* of the life and times of the central hero of the novel, Wokulski. They also enable the reader to view the contemporary reality from different angles, through the experiences and attitudes of Wokulski himself, and then through the eyes of a subordinate character (Rzecki) who represents another dimension of the Romantic generation of which Wokulski himself is a product. Ideologically, Prus through the memoirs exposes the utter inability of the Romantic *Weltanschauung* to come to grips with the problems of reality in the post-1863 period. Rzecki's naively stubborn and amusing Napoleonism clearly demonstrates this. His worship of Napoleon has its roots in his father's blind faith in the Emperor, which he inculcated in his son and which involves not only an adoration of Napoleon I but of the entire Bonaparte dynasty (something Mickiewicz was almost always careful to keep separate, as we can see in the distinctions he draws between Napoleonism and Bonapartism in the articles he wrote in the 1850's for the socialist paper, *La Tribune des peuples*). Describing the home of his aunt, with whom he and his father lived after the death of his mother, Rzecki recalls early in *The Doll* that:

My aunt's walls hung with pictures of the saints, but as many of them as there were they did not equal in number the Napoleons with which my father adorned his room. There was one Napoleon in Egypt, another at Wagram, a third at Austerlitz, a fourth at Moscow, a fifth on the day of the coronation, a sixth in apotheosis. When my aunt, horrified by so many secular images, began to hang a brass crucifix on the wall, father, in order — as he said — not to insult Napoleon, bought himself a bronze bust of the Emperor and also hung it up over his bed.

"You'll see, unbeliever", my aunt often lamented, "for these tricks they'll bathe you in pitch."

"Not at all! ... The Emperor will keep me from any harm", answered my father. (Vol I, Chap. III)

Rzecki's hopes for the future of Europe and (of course) ultimately Poland rest with "little Lulu",[9] who he is convinced will become French emperor under the name of Napoleon IV. As Emperor, Lulu will defeat the Germans, will bring justice again to the entire world, as Rzecki says his father had foretold. Rzecki's hopes in this regard, however, are doomed to frustration, but his faith is so strong it never can be utterly defeated. At one time he was certain that Marshall Mac-Mahon would attempt a coup on behalf of the young Napoleon. But then Mac-Mahon fell and Grévy became French president. Napoleon (IV) went off instead to learn the art of war fighting along with the English in Africa:

In half a year (Rzecki comforts himself) he will come back covered with glory, so much so that the French themselves of their own accord will pull him back and raise him to the throne ... Long live France and the Napoleons! (Vol. III, Chap. III)

Little Napoleon, Lulu, is serving in the English army and is fighting the Zulus in Africa just as his grandfather Napoleon the Great did. Then when the English finish the war, they'll make Napoleon a general and will tell the French:

Dear fellows! Here you have a Bonaparte who fought in Africa and covered himself with immortal praise just like his grandfather. Make him your Emperor as you made his grandfather Emperor, and for that we'll get Alsace and Lorraine for you from the Germans. You'll pay them a few million, but that's better after all than starting a new war, which would cost certainly more than ten million and would be doubtful for you ...

The French naturally will make Lulu Emperor, will recover their territory, will pay what they have to, will enter into an alliance with the Germans, and then Bismarck having so much money will show his stuff! (Vol. III, Chap. III)

Rzecki's romantic dream of another Napoleon on the throne of France is shattered by the rumor that Lulu died in Africa, "and to make matters worse in battle with a savage people of whom nobody knows where they live or how they're called, since after all no nation can call itself Zulu". Before the novel closes Rzecki contributes five rubles towards an office for the dead for *Le Prince Impérial*, but, as he says, "only for an *intention*, since it's possible he didn't really die, though everyone says he did".

[9] Napoléon-Eugène-Louis-Jean-Joseph Bonaparte (1856-1879), the only son of Napoleon III and the Empress Eugénie. He was known popularly as "Le Prince Impérial".

Prus's *The Doll* with its unconcealed anti-romanticism and anti-Napoleonism was published serially in the Warsaw paper *Kurier codzienny* (Daily Courier) between 1887 and 1889. It appeared in book form in three volumes for the first time in 1890. This period, beginning in the 1880's and continuing into the early years of the twentieth century, witnessed an impressive growth of Polish historical fiction. Henryk Sienkiewicz's well known *Trilogy*, *Ogniem i mieczem* (With Fire and Sword), *Potop* (The Deluge), *Pan Wołodyjowski* (Pan Michael), appeared from 1883 to 1888. *Quo Vadis*, which was to establish Sienkiewicz's international reputation, came out in 1896. Prus's novel of ancient Egypt, *Faraon* (The Pharoah), his only major work to win recognition outside of Poland, was published the same year. The historical plays of Wyspiański and the novels of Żeromski and Reymont were to follow in rapid succession.

The rise of this historical fiction was the result, in part, of the rebirth of Polish historical science in the last three decades of the nineteenth century. Conditions in the Austrian province of Galicia played a central role in this development. By 1867, the province had achieved an autonomous, self-governing status within the Austro-Hungarian Empire. Polish intellectual and cultural life began to flourish anew. The ancient University of Cracow again became an important center of Polish learning, the University of Lwów rose to prominence as the second leading center of learning in Galicia, and in 1870 a research academy was founded in Cracow which attracted scholarly and scientific talent from not only Galicia but other parts of partitioned Poland.

In this new liberal climate of Galicia in the late 1860's, 1870's, and lasting through World War I and the restoration of Polish independence, it became possible to undertake the systematic evaluation of the history of Poland, particularly in an effort to determine the causes behind the collapse of the once great Commonwealth. The historians grouped in Cracow, notably Józef Szujski (1835) and Michał Bobrzyński (1849-1936), were of conservative orientation and were associated with the dominant political organization that sprang up in Galicia after 1860 — the Conservative Party. These Cracovian conservatives, whose ranks included prominent members of the Galician intellectual elite, were known as the *Stańczyks*[10] from a political pamphlet — *Teka Stańczyka*, Stańczyk's Portfolio — drawn up chiefly by the historians but published anonymously in 1869. Their desire to safeguard the newly- and hard-won self-governing status of Galicia and to ensure that nothing would be done to jeopardize the privileges that had finally been granted led the Conservatives to a policy of close cooperation with the Austrian imperial authorities.

In their assessment of the past, the Cracow historians (whose scholarly output was considerable) tended to view the downfall of the old Polish state as the

[10] The name is derived from Stańczyk, the court jester of King Zygmunt I (1506-1548).

direct result of internal weaknesses. They emphasized in particular the anarchy which for them was the consequence of a weak central authority. There was little sympathy on their part for the traditional Polish attitude that Poland fell because it happened to have powerful and avaricious neighbors. Given such an approach to Polish history, the Cracow historians could hardly have been expected to look with favor on the Romantics' messianic ideology or their nostalgic and sympathetic evocation of the Polish Saxon-Sarmatian gentry past where they themselves placed most of the blame for Poland's political woes.

Opposition to the "revisionism" of the Cracow historians came from two sources: the Lvovian historical school consisting principally of Ludwik Kubala (1838-1918), on whose research in seventeenth century Polish history Sienkiewicz drew heavily for his *Trilogy*, Oswald Balzer (1858-1933), and later Szymon Askenazy (1867-1935), and the Warsaw historians Tadeusz Korzon (1839-1918) and Władysław Smoleński (1851-1926) who sought to draw attention, unlike the Stańczyks, to the constructive achievements of the old Polish state. The tenor of this opposition to the conservatism of the Cracovians was perfectly captured by the Positivist poet, Adam Asnyk, in his poem "The New Historical School" (Nowa szkoła historyczna, 1890):

The new school of history / Having acquired an accurate method of research, / Brought in lump-form from Germany / And spread now beyond the Vistula, / As well as a fresh outlook, / Studies history with a new measure. / Not wanting to fall into the old errors / Of foggy messianic day-dreams, / It wound its halter / Around stiff Polish necks / In order to bend / The savage impulse of life / Before the logic of accomplished facts / And the clever symbols of victories. / It did not notice that the worship of power / Led it too far / And that from its hands the partition of Poland / Took the cross of apostolic mission. / In the full light of its investigations / Despotism shines like a clear star / And the patriotism of the Targowica / Rises purely on high ... / Since Kościuszko was a madman / Who roused the proletariat! / And so on ... and so on ... / It spins still bolder conclusions, / And destroying old judgments / It will not restrain its flight / Until it proves that King Herod / Was a benefactor to orphans.

An important and characteristic feature of this extensive Polish historical research and writing was the almost exclusive preoccupation with the history of the independent Polish Commonwealth. Until the early twentieth century, the Polish historians generally refrained fom dealing with the post-Partition period. In its own development, Polish historical fiction followed the same path. The works of the first important Polish historical novelist, Józef Ignacy Kraszewski (1812-1887), covered the whole range of Polish history from the prehistorical period (*Stara baśń* [An Old Fable] 1876) to the time of Stanisław August Poniatowski (second half of the eighteenth century). His most impressive effort remains two novels of the cycle devoted to the Saxon period — 1697-1764 — when kings of the House of Saxony occupied the Polish throne: *Hrabina Cosel* preach from the podium of a lecture-hall in the Collège de France. For Żeromski, more than a half century later, the Napoleonic legend took on a very different appearance, one that had little in common with that of the Romantics. The novelist grew up and was educated in Russian Poland during the Positivist anti-Romantic reaction. His views in part were shaped by positivism; *Ashes* reveals this no less than his great social novels where society emerges more significant than the individual and where individual self-sacrificing heroism is incapable by itself of eradicating social evil. Much has already been made of Tolstoyan influence on Żeromski, but *Ashes* is rarely approached in terms of an antithetical response to the Napoleonism of *Pan Tadeusz*. It is in such terms that we shall view it now.

Mickiewicz, we saw, refrained from portraying war and military scenes in *Pan Tadeusz*. These are reported indirectly, with an absence of detail, and in the spirit of myth. Żeromski, however, has concentrated on war and military conflict. But war for Żeromski is not the vital inspirational force that it is for Mickiewicz, or the joyful, elemental release that characterizes its treatment in Sienkiewicz. It is ghastly, ugly, miserable. In *Pan Tadeusz* the Napoleonic campaigns linger in the wings and emerge on stage only late in the work and then in a ceremonial way. In *Ashes*, they come to dominate much of the action of the novel. For Mickiewicz Napoleon is a god-like messianic figure; for Żeromski, he is a remote, emotionless leader whose idealism has been consumed in the fires of personal ambition. In *Pan Tadeusz* the Polish Napoleonic alliance is seen as something grand and glorious. In *Ashes*, on the contrary, it is exposed as inglorious, even senseless folly.

Żeromski's attitude toward the Napoleonic period in Polish history was obviously conditioned by Positivist teaching. But *Ashes* cannot be situated ideologically solely within the framework of positivism. Żeromski's stance is decidedly anti-militarist. War is portrayed in all its horror; there is even an emphasis on its more macabre aspects, as if the author wished to drive his point home with maximum force. Yet the vision of the Polish Napoleonic adventure that unfolds in *Ashes* is not entirely negative. From the misery of war and the despair of frustrated hopes, Żeromski tries to find something redeeming, something noble in the Polish contribution of blood and tears to the Napoleonic campaigns: the desire for liberty and national dignity that is so strong it can reconcile its idealism with humiliation and ignominy if only its efforts ultimately serve the national cause. This is the positive statement that Żeromski succeeds in making in *Ashes*, but his indictment of the Napoleonic delusion carries a greater impact.

In his repudiation of the glorification of the Napoleonic adventure, Żeromski concentrated on two episodes in *Ashes*: the Santo Domingo expedition of

1802-1803 and the Spanish campaign of 1808. We are not hard pressed to discover why. These episodes stand out most vividly as the best examples of the callousness with which Napoleon approached the use of his Polish troops. In both instances, Poles were brought in to crush national liberation movements. The irony here is inescapable. Poles who had rallied to Napoleon's cause as the great remaining hope for Poland's liberation from oppression found themselves serving in campaigns expressly undertaken by Napoleon for the purpose of destroying native revolts which demanded freedom from foreign (i.e. French) domination.

The expedition to Santo Domingo had come about for just such a reason. The famous native leader Toussaint-Louverture (1743-1803) had requested Napoleon's approval of a form of constitutional government for the island as a French colony. The Emperor refused, but with vague promises succeeded in getting Louverture to lay down his arms. By a ruse, the French captured the Negro leader and deported him to Paris, where he was promptly imprisoned in June, 1802. Ten months later he was dead of consumption. After his death, the natives on the island renewed the struggle, stirred also by reports that the French were planning to reintroduce slavery in their colonies. Captain-General Leclerc, the first husband of Pauline Bonaparte, was entrusted with the task of quelling the revolt. It proved to be a costly assignment that ended in defeat and humiliation.

The lengthy description of the Santo Domingo campaign in *Ashes* is couched in the personal narration of an old Polish veteran who, dressed as a beggar, on crutches, comes to beg a night's lodging in the manor where the principal protagonists of the novel, Raphael Olbromski and Christopher Cedro, are staying. The veteran is slowly making his way home from the Battle of Austerlitz, where he lost a leg. He is, in his appearance and manner, one of the "beggar-veterans" Mickiewicz describes near the end of Book I in *Pan Tadeusz*:

And yet now and then, like a stone from the sky, news came even to Lithuania; now and then an old man, lacking a hand or a foot, who was begging his bread, would stand and cast cautious eyes around, when he had received alms. If he saw no Russian soldiers in the yard, or Jewish caps, or red collars, then he would confess who he was: he was a member of the Polish legions, and was bringing back his old bones to that fatherland which he could no longer defend. Then how all the family — how even the servants embraced him choking with tears! He would seat himself at the board and tell of history more strange than fable; he would relate how General Dambrowski was making efforts to penetrate from the Italian land into Poland, how he was gathering his countrymen on the plains of Lombardy; how Kniazewicz was issuing commands from the Roman Capitol, and how, as victor, he had cast in the eyes of the French a hundred bloody standards torn from the descendants of the Caesars; how Jabłonowski had reached the land where the pepper grows and where sugar is produced, and where in eternal spring bloom fragrant woods; with the legion of the Danube there the Polish general smites the Negroes, but sighs for his native soil.[12]

This "land where the pepper grows and where sugar is produced" is, of course Santo Domingo, but consistent with the whole tone of *Pan Tadeusz*, there is no suggestion of any of the horrors the Poles experienced in the campaign there. The Poles instead are victors, they are "smiting Negroes" in this lovely land where "in eternal spring bloom fragrant woods". It is a very different Santo Domingo that emerges from the pages of Żeromski's *Ashes*. The veteran recalls that in June 1802 as a member of the 113th demi-brigade of the Polish Legion of the Danube he boarded a vessel in Livorno, Italy, for the Atlantic crossing. A storm forced the ship to pull in at Malaga, Spain. In July it set out anew; after a brief stop in Cadiz, it sailed again in August and reached the Antilles in October.

For the old legionnaire the island was no place of beauty, but of unspeakable torment physically and spiritually. The weather was oppressive, yellow fever ran rampant, circumstances caused treachery and inhumanity that rapidly tore the Franco-Polish forces to shreds, and all that remained was chaos and degradation:

Human tongue cannot describe the disorder which finally overcame the French army. Scarce a third of the soldiery kept on their feet. Everyone discouraged, sad, in worn-out clothing, looked for the place where death would strike him. And so some threw themselves into the maddest lewdness, into license and outlawry which the mind could scarce conceive. They revelled, drank, killed with their bayonet, raped property and virtue, spent their nights with creole, negro, and mulatto women, while others prepared for death in fasting and spent their nights lying face downward on the ground beside their beds. The hospitals were full of sick, lying in swaths on the floor, without care, help, or attention. All subordination ceased, military regimen vanished completely. The common soldier was equal to the general. No one now occupied himself with thought of winning battles, of fame. The hearts grew icy, bandit souls declared themselves. The bayonet became the ultimate law. Everyone was only basely, ignominiously afraid or even more ignominiously debauched.[13]

In a few weeks time, the veteran recalls that of his battalion which numbered a thousand men, only a hundred and some odd were left! All told, of the 6,000 Poles who landed in Santo Domingo, 300 men and less than a score of officers survived.[14]

The legionnaire himself was felled by yellow fever. After his discharge he returned to Europe, rejoined the French army, and saw action at Austerlitz, where the campaign cost him a limb. Stephen Trepka, who has been listening

[12] *Pan Tadeusz*, pp. 30-31.

[13] Stefan Żeromski, *Ashes*, II, trans. Helen Stankiewicz Zand (New York and London, 1928), 442.

[14] This is the figure given by Marian Kukiel in *Dzieje Polski porozbiorowej 1795-1921* (London, 1961), p. 72.

to the veteran's tale together with Olbromski and Cedro, interrupts him near the end of his narrative to accuse: "At the beck of a usurper, to foster the intrigues of his co-mountebanks, you trampled free peoples, wiped out tribes ..." And the veteran's reply?

Vengeance is mine, saith the Lord! Pass no judgement, my friend. For only look what has happened: for twelve years we spilled our blood on the outskirts of the world. My brothers almost to a man went to their graves in mortal despair. And now! The German Empire scattered like a pile of chaff. We helped at Jena, at Auerstadt, and the Prussian nation is overthrown? Usurper! This same usurper is master of Berlin and master of Vienna. You don't know what you are saying ... Long live the Emperor! A hundred times! A thousand times! Eternal glory to him![15]

For all the suffering he has been made to endure, for all the misery his pursuit of the Napoleonic dream has brought him, the old soldier — like so many Poles of his generation — refuses to surrender to remorse or disillusionment. Admission of the ignominy of the Santo Domingo expedition is an act of self-betrayal he cannot permit himself to make. And in this Żeromski sees the power — and pathos — of the Napoleonic myth.

The fierce Spanish campaign of 1808 by which Napoleon dethroned Ferdinand VII and placed his brother Joseph on the throne is treated in similar fashion, though in far greater and more vivid detail. The high point of the campaign was the bitter siege of Saragossa, in which Polish troops played a decisive role. In *Ashes* the siege functions as one of the main supports of the artistic and ideological structure of the novel. Nowhere else does the author so nakedly strip the Polish Napoleonic myth of its aura of glory; nowhere else does style come so perfectly into the service of ideology.

The campaign is seen through the eyes of Christopher Cedro, who like his distant cousin and friend Raphael Olbromski joined the Polish forces fighting under Napoleon not long after their enthusiasm was enflamed by the old legionnaire's tale of the Santo Domingo expedition. Cedro and Olbromski split up in Łowicz, Cedro joining the Northern Legion under General Zajączek, thereafter serving in Spain with the Legion of the Vistula, Olbromski going on to participate in the campaign in Galicia (1809) under the command of Prince Poniatowski.

In his account of the Spanish campaign Żeromski has focused on certain actions connected with the siege of Saragossa that carry the burden of a symbolization of the Polish presence in Spain. The first of these is the attack on priests in a barricaded house on Sant' Engracia street where much of the fighting takes place. The seizure of the house is celebrated with a mass rape of a group of girls the French and Polish troops find huddled together in a room. Cedro's

[15] *Ashes*, p. 447.

pursuit of one of the girls leads him to another room where in a grotesque scene he is overpowered by a group of older women intent on slaughtering him. He is saved by some *voltigeurs* who tear the women away from him and then pitch them out a window onto a courtyard two floors below. In the next action, the convent of the Virgins of Jerusalem is overrun and Cedro bears witness to the humiliation of a group of nuns forced to dance naked to the accompaniment of soldiers beating on pots and pans. The shock of the scene is reinforced when another officer shows him the body of a young nun who plunged a dagger into herself when attacked by a half dozen soldiers. The common element in these encounters is defilement; and if we pass from the grimness of the reality to the realm of symbol we find here, I think, an implication of Napoleon's defilement of the purity of Polish ideals and ambitions. It was the awareness of this betrayal that prompted one of the legionnaire poets who participated in the Spanish campaign, Kantorbery Tymowski, to express his disenchantment in the poem "Dumania żołnierza polskiego w starożytnym zamku maurów nad Tagiem" (Thoughts of a Polish Soldier in the Ancient Fortress of the Moors over the Tagus):

Buttressing the foundations of Iberian thrones,
I lose the honor of defending my native fields! ...[16]

The incidents we have described above are but a prelude to the most powerful episode of the siege of Saragossa and the entire Spanish campaign: the battle waged within an insane asylum. Here Żeromski has found the perfect symbol for the madness of the Polish Napoleonic cult that has led to humiliation and spiritual sickness first in Santo Domingo and again in Spain. French and Polish troops pursue Spanish partisans into a building they are soon to discover is an insane asylum. In order to erect protective barriers between themselves and their enemy both sides release as many inmates as they can. When the inmates come streaming out of their cells screaming and rushing in all directions, the scene takes on a macabre, nightmarish aspect:

There now burst from the corridor, like leaves driven by the wind, dancers, declaimers, singers, orators, thoughtful men, indifferent men, men blind with frenzy, likenesses of dogs crouching for a spring, and likenesses of felled trees which a fungous rust of indifference seemed to have covered and consumed, men without faces and with frenzied eyes, others with hideous faces which had no eyes, horrible chimeras in women's forms, frightful creatures with the eyes of wolves and tritons, with chains on their arms and in jackets with the sleeves tied in the back. The mob came out to meet the soldiers and barred their way. The roar of beasts, the voices of storms, the groans of the wind in a sylvan wilderness and the night song of the sea lashed into a frenzy by the new moon, the cry of the deepest pain of birds and the laughter of happiness evoked

[16] In *Zbiór poetów polskich XIX w.*, ed. by Paweł Hertz (Warsaw, 1959), I, 462.

from nothingness by a musical instrument, tears over an empty cradle and the euphonic hymn of the soul gazing into the open heaven — all this burst at the newcomers from this multitude.[17]

Żeromski intensifies the impact of the scene by concentrating now on a single act of insane savagery played out before Cedro's eyes:

When Wyganowski came up the stairs with Cedro, he saw before him, in the coils and clouds of powder-smoke, two men in rags, with white skulls, who had seized each other by the throat and had plunged their teeth into each other's flesh. They had just fallen to the ground. Now one, now the other was on top. They were tearing each other with their jaws like mad dogs. Their naked arms, knees, thighs, bellies, shoulders, necks, flew about, trembling and quivering. They tore the living blood from each other with their teeth, they pressed each other with their knees, ripped each other with the sharp points of their claws; they clung to each other so frightfully that they seemed to be one body with two heads, many arms and legs. The blows of their fists doubled, tripled, multiplied a hundredfold. They tortured each other, striking skull against skull with a cacking giggle of venomous malignity. One heard a cracking of bones and a throaty rattle, a cracking of bones and a rattle ... At length one of them remained on top for a longer while. The other rattled beneath him. Only his head still strained to lift itself, and the neck continued to brace itself for a blow. But the strangler would no longer surrender his advantage. His bloody frenzy did not cease for one moment, even when the white, bruised skull of the victim drooped powerlessly into a pool of blood. He sucked his gushing wounds, raised the lids of his eyes, and stared into them, looked to the bottom; he watched with his hands for each of the man's last breaths and caught them instantly, while they were still in the windpipe. At length he dealt the man a final blow. He spat for the last time into the powerless mouth. He rose. He passed the strange meat of his eyes, the hell of his leer, over the row of astounded soldiers. He saw them now. A laugh broke from his lips, a howl, a sob, a titter ... He lifted his shoulders and like a joyous lion leaped from above into the midst of the crowd. He caught by the beard the officer walking in the third file, the soldier next to him by the throat, and, roaring with bliss, his bared teeth foaming with a froth of happiness, died on the pointed bayonets ...[18]

In striking contrast to the lyricism with which *Ashes* (like most of his works) is infused, Żeromski has called into use the stylistics of naturalism with its emphasis on the macabre, its passion for the detail of horror, its love of blood and the blood-red death-black play of color in order to portray as graphically as possible the insanity of war. It was a technique the novelist had demonstrated a preference for earlier in his stories on the Insurrection of 1863, *Rozdziobią nas kruki, wrony* (Ravens and Crows Will Tear Us to Pieces, 1895), and which he was to employ again, no less effectively, in his novel on the same subject, *Wierna rzeka* (The Faithful River, 1913).

Like the old soldier who told him of the campaign in Santo Domingo, Cedro's

[17] *Ashes*, p. 608.
[18] *Ibid.*, p. 607.

faith remains unbroken through all his experiences. It falters once in a while, certainly, but never gives way. During the siege of Saragossa — at the height of the campaign, when the Poles could not have missed the bitter irony of their own situation — Cedro's faith is tested by his fellow officer Captain Wyganowski:

"What in the devil's name are we doing here? How long are you thinking of staying here?" he asks Cedro.

"Until they tell me to go somewhere else. That's clear, surely. This war must end some time."

"And for whom are you fighting?" the Captain blurted out with bloody sarcasm.

At this question Cedro leaped at Wyganowski crying: "I am fighting with all my soul for my country!"

"Very well, continue to speak that slogan before yourself and God. I don't wish any longer. I am going."[19]

Cedro's almost blind, unquestioning belief in the rightness of Napoleon's cause contrasts with the attitude of his lifetime friend Olbromski. With the end of the campaign in central Poland in 1809, Olbromski returns to the estate of Wyrwy bequeathed him by his uncle, Squire Nardzewski. Until 1812, the year of the campaign against Russia, all his energies are devoted to the productive management of his estate and the cultivation of the land. Then, when the campaign against Russia is about to be launched and the flower of Polish youth is again rallying to Napoleon, Olbromski is visited one day, to his surprise, by Cedro on his way to join the Polish forces. Unable to restrain his excitement Cedro urges his friend to leave with him. Older and wiser now, Olbromski is reluctant to leave his estate and the land he is cultivating for the folly of war a second time in his life. He sees a greater need in the construction of the house he has so long been planning. Cedro's response to this unheroic attitude is one of almost utter incredulity:

"You are putting up a house!" Cedro burst out in such merry and sounding laughter that Raphael flared up. But shame overcame him at the same time. "Well, what do you think?" he grumbled. "I took over the estate and I had to get to work some time ... I can't dally forever."

"You are beginning to work when all of us are starting out to war? Seventy thousand of ours have set out ..." Olbromski almost wept. All at once an inexpressible regret came over him at the thought of these fields, these newly cleared lands, these fences. He looked at the house gleaming white in the distance, amid the spreading trees.

Cedro's enthusiasm is too great to be denied, however:

"And so you are going to war with me!" Cedro said suddenly, changing the subject of their conversation. "It will be a great war. A war of wars."[20]

[19] *Ibid.*, pp. 743-744.

[20] *Ibid.*, pp. 755-756.

With a curse on his lips, a curse perhaps of resignation ("I go too, damn it!") Olbromski agrees to join Cedro.

There can be little doubt here as to Żeromski's Positivist orientation. The author's sympathies are wholly with Olbromski. Returned from war, his pursuits are those of peace: he is managing an estate, he is building, he is making the soil produce. His contribution can be measured in positive terms. It may be small, it may lack the glory and brilliance of another campaign under the flags of the Emperor, but its results are bound to be more enduring. He surrenders finally to the romantic intoxication of Cedro not out of any sincere conviction but out of a sense of futility at trying to convert him to the sense of his own outlook. The triumph here is the triumph of the bonds of friendship, nothing more.

The narcotic power of the Napoleonic myth is strikingly dramatized by Żeromski in the shift in character of the principals that we can observe at this point. The former impulsive, adventuresome, and heroic nature of Olbromski has given way under the impress of war to a new maturity and sobriety. When we see him at the end of the novel he is tired of strife and romantic heroics and yearns for the satisfaction of constructive accomplishment. The more submissive and delicate Cedro, on the other hand, has succumbed completely to the spell of the Napoleonic mystique and exults in the thought of the glory of new wars in the Emperor's name. The values Olbromski has come to embrace are meaningless to him; on the eve of the campaign against Russia his enthusiasm is so high it is impossible for his view to embrace anything but uniforms, banners, and triumphs. He no longer even speaks of the goals of the campaign. In his consciousness it has become a thing unto itself.

Żeromski's unsympathetic treatment of the Napoleonic campaigns is reflected in his portrayal of the Emperor himself. There is not the "humanization" of the Hero as in *War and Peace* or the memoirs (*Trzy po trzy*, Topsy Turvy) of the Polish comedy writer Alexander Fredro, who served with the French in the Russian campaign. Napoleon is not a ludicrous figure in *Ashes*. Rather he is an unenviable human being in whom the flame of idealism has flickered out and who has become the victim of his own ambition.

Napoleon appears on two occasions in *Ashes*. The first time occurs during the campaign in Spain. Cedro has been wounded in action at Burviedro. Lying on a stretcher on the field, he is almost asleep when he hears soldiers shouting the name of the Emperor. In a bitterly ironical scene, a proclamation of Napoleon is read. It declares that the Emperor has "abolished and forever annulled the Holy Inquisition, released its prisoners, and quashed its persecutions, that he reduced the number of monasteries and convents by two thirds, that he overthrew and forever cancelled the ancient rights of feudal lords, that he abrogated and destroyed all privileges." Almost in a delirium, Cedro says to himself:

Now you know why we stormed old Saragossa, Aljafaría with its prisoners, why we drowned our lances in the blood of dark masses at Tudela. Your bill of rights is written in our blood, O Spanish prisoner![21]

At this point the Emperor himself comes to Cedro and is seen as though in a vision. His face is "pale and mysterious, verily like the moon hidden in clouds. Blurred eyes dart in it, dart in all directions, then retreat once more into their shoulders to watch like lions in ambush ..." Christopher sits up on the stretcher frightfully pale, drenched with perspiration, his mouth full of blood. His eyes plunge into the figure of the approaching man "like greedy fangs". They root him to the spot. He stops.

"Sire!" said Cedro.
The dark martial eye of the commander clashed with the eyes of Christopher. The calm face, as if carved in an unknown metal, was turned to him with stern interrogation.

In a voice "cold and toneless" Napoleon asks him what his wish is. Cedro identifies himself as a Polish soldier, a lancer from Tudela. He does not give his name but says:

I left my father's house — I believed that it was for my country and now — on foreign soil — say that it was not in vain, that it was for my land, — Emperor, Emperor!
The mute and deaf eyes plunged into the delirious eyes of the invalid, eyes mad with deathless love. Moveless and thoughtful stood Napoleon. Who knows? Perhaps in those inspired eyes he saw his own young soul. Perhaps the rosy snows of the crags of Monte Oro, perhaps the pinions on the crests of Monte Rotondo, perhaps the stony shores of the island lashed by the foams of the raging sea. Perhaps he weighed, for a moment, his Corsican love of liberty with the crown of ruler over peoples strange to him and the sceptre of Charlemagne. Perhaps he sighed in sorrow over that which had already died in his own soul, crumbled, and been scattered by the winds like the stalks of a dead flower, over the engrossment of a young, just, and proud soul in the misfortunes of its country.

"Vive la Pologne!" Cedro sought to cry out, as, strengthless, he fell back upon his mattress. He did not cry, however, for the words were drowned in a stream of blood which suddenly gushed from his mouth.

The Emperor stood over him for a long moment. He looked into his face with stony eyes. Finally he raised his hand to his hat and said: "Soit". He walked away in a slow, cold, measured step. Behind him a throng of generals. He disappeared between the columns of infantry, between the massed files of horses.

In the weakness of his illness, Cedro experiences a rare moment of doubt and seeks some affirmation of the Emperor's dedication to the Polish cause. Such an affirmation would make tolerable the participation of Polish forces in the Spanish campaign. The sphinxlike figure of Napoleon remains silent, however.

[21] For this and immediately following quotations, *ibid.*, pp. 633-635.

The blood gushing from Cedro's mouth becomes a vision of the thousands of Polish troops who laid down their lives fighting for the Emperor in the belief that it was for the good of their country. The "Soit" that falls feebly from Napoleon's lips before he takes leave of so oppressive a sight is the measure of his incapacity to feel.

Napoleon appears for the second and last time in the novel when at the end of the work Cedro and Olbromski, again in uniform, participate in a review of the army at Orsza with the thousands of Polish troops under the command of Prince Joseph Poniatowski united to the *Grande Armée*. The Emperor himself appears to take the review. Cedro, looking at the figure of Napoleon, cannot take his eyes from him:

For, lo, he sees, in waking life, the realization of his great dream. The Emperor has kept the word given at Madrid to the weakest of his soldiers, to the invalid dying in the field. For that one word he has united regiments, uniformed and fed them, and set them in motion. He has drawn foreign nations to his side ...[22]

Żeromski's description of Napoleon departs here only slightly from the earlier image of emotional numbness and spiritual emptiness. In the crowd of massed troops ranged before him the Emperor spots — and recognizes — Cedro. The young man's enraptured expression for a fleeting instant reminds him of the idealism that once, a long time past, stirred him too, and for the only time in the novel a suggestion of a smile tinged with sadness by the realization of his spiritual predicament breaks through the rigid immobility of his face:

His face was cold and sullenly indifferent, like a slab of stone. The look of the eyes passed face after face, moved over the eyes of thousands as over a lifeless trail. The dreadful eyes in which the human throngs had learned to see only joy and anger were at this moment neutral, indifferent, veiled with the vastness of distant thoughts.

His hand pulled the reins of the horse, and a command fell from his lips. The Emperor stopped. He raised his eyes. He looked into the regiment. He saw every face, he passed in turn over every one. He met the eyes riveted upon him, of Christopher Cedro, eyes turned to stone with soldierly fealty, eyes sworn to undying faith. For a moment something glimmered in the hard orbs of the Emperor, like a distant gleam in a grey, formless, featureless cloud. Memory became a clear vision of reality. A fleeting, half-sad smile passed over the granite face.

The smile leaves as quickly as it appears and the Emperor's face again becomes a slab of stone, of granite suggesting, one is tempted to remark, the barren island of St. Helena, where he was soon to spend the last days of his life and where, surrounded by a handful of supporters who remained loyal to him to the very end, he was to begin himself the legend that conquered Europe a second time in the name of Napoleon. Some would say it was the Emperor's greatest conquest.

[22] *Ibid.*, pp. 757-758.

DOSTOEVSKIJ THE HUMORIST: 1846-1849

VICTOR TERRAS

University of Wisconsin

The critics of the 1840's whose image of the young Dostoevskij was not biased, as is that of later critics and ours, by the great novels of the mature writer did not think much of Dostoevskij's intellect. Belinskij, for instance, believed that Dostoevskij had created "the first Russian social novel [*Poor Folk*] ... as artists usually do, i.e., not suspecting himself what was flowing from his pen".[1] His exclamation to the young writer himself expressed the same idea even more bluntly: "You don't understand, do you, what you've written there!"[2] Of course Dostoevskij knew what he had written and it was Belinskij, the mature critic, and not the beginning writer who was being naive. Annenkov, a clever practical psychologist, has left us an insightful description of Dostoevskij listening to Belinskij's fatherly advice — with respect, but noncommittally and without genuine interest.[3] Yet there are aspects of the young Dostoevskij's work which Belinskij and his contemporaries may have understood better than later generations of critics and scholars. Could anybody whose first acquaintance with Dostoevskij had been made through *Crime and Punishment* or *The Brothers Karamazov* list the writer's humor as one of his principal assets? Belinskij does just that — one of the great critic's few insights into Dostoevskij which, I think, really do credit to the genius of both.

Its humor is one of the best things about *Poor Folk*, as Belinskij immediately recognized. It is also essential to the basic theme of the novel: a challenge to and a refutation of Gogol's *Overcoat*. Devuškin, in order to both resemble and transcend his prototype Bašmačkin, must be a genuinely ridiculous person. He is. The rules of the ἄγων stipulate that Dostoevskij's story be as sad, but also as funny, as Gogol's. It is. In fact, Makar Alekseevič Devuškin, very much in contrast to the dull Akakij Akak'evič, has himself a sense of humor and the

[1] According to P. V. Annenkov, *Literaturnye vospominanija* (Moscow: Gosizdat, 1960), p. 282.

[2] According to Orest Miller, "Materialy dlja žizneopisanija F. M. Dostoevskogo", *Biografija, pis'ma i zametki iz zapisnoj knižki F. M. Dostoevskogo* (St. Petersburg: A. S. Suvorin, 1883), pp. 60-61.

[3] See Annenkov, *op. cit.*, p. 283.

whole novel, while presenting an ampler view of life than *The Overcoat*, is also richer in the variety and depth of its humorous effects.

The point of it all is, of course, that in spite of being so funny the hero of *Poor Folk* is still a real person, not a puppet on a string as is Akakij Akak'evič. Combined with the tragic pathos of the lyrical and the rhetoric of the "rebellious" passages, the comic ethos which never disappears from Devuškin's diction right to the last, desperate effusion makes the little clerk a so much more deeply human figure.

The principal mode of humor in *Poor Folk*, although perhaps not the most readily identified, is that of travesty.[4] Devuškin is making a travesty of his role as a lover in the sentimental tradition, and Dostoevskij is making a travesty of the sentimental *roman d'amour*. Everything seems to prove that Makar Alekseevič just wasn't born to be a sentimental lover. Everything in the novel seems to "work out" so that Devuškin's and Varen'ka's romance is turned into a farce.

Devuškin's courtship is hyperbolically platonic. So persistently does he plead his parental obligation to Varen'ka (it is based on a consanguinity which is not only distant but also imaginary), his disinterested sympathy, and his "fatherly feeling", that his chances of ever getting a second kiss from her (one is mentioned in the very first letter!) are soon reduced to nil. He sells his uniform to buy Varen'ka flowers, candy, and grapes (in Petersburg, in the month of May!) — because "they are good for a convalescent and the doctor recommends them for quenching the thirst, just simply for thirst" (I, 94).[5] As fits a sentimental lover, Devuškin grows very thin as the courtship goes on — but for the prosaic reason that he cannot afford both to eat well and buy Varen'ka expensive trinkets. Besides: "Who is it that says I've grown thin? It's slander, nothing but slander again! I am well and hearty and getting so fat that I'm quite ashamed of myself!" (I, 95).

In spite of the extravagant innocence of Devuškin's "affair" with Varen'ka neighbors soon start wagging their tongues. The way *they* see it, the travesty turns into a coarse burlesque:

> ... but my landlady, did she raise the roof again! Well, I wouldn't have minded that. Let that worthless woman scream all she wants, but, for one thing, it was the disgrace of it and, for another, God knows how, but she had found out about our liaison and was shouting such things about it at the top of her voice that I was left numb with horror and closed my ears with my hands not to hear it. But the thing is that the other roomers didn't close theirs but, quite on the contrary, pricked them up. (I, 151)

It seems that Devuškin uses "liaison" (*svjaz'*) as a euphemism, achieving of course the contrary of the desired effect because, in Russian as in English, the

[4] Cf. my article "Problems of Human Existence in the Works of the Young Dostoevsky", *Slavic Review*, XXIII (1964), 79-91.

[5] All quotations are translated from F. M. Dostoevskij, *Sobranie sočinenij* (Moscow, 1956-58).

word is suggestive of an intimate relationship — which Devuškin does not know! In the Russian original all further references to the "liaison" are in the pronoun of the third person, as he is trying to avoid another direct mention of the scandalous accusation. Naturally, the resulting awkward syntax, rather than concealing "it", makes "it" even more conspicuous. The whole passage is most typical of the naïveté with which Makar Alekseevič tends to repeat injurious statements made by others about him (or Varen'ka). There also may be a trace of voluntary buffoonery about the present passage.

Devuškin's literary friend Ratazjaev, in the meantime, finds the whole affair positively hilarious, much to Makar Alekseevič's dismay. In his letter of August 11 (which, incidentally starts with the words "I am lost, we are both lost ..." — echoing Julie's "tout est perdu ...")[6] Devuškin reports that one of his letters has been intercepted by Ratazjaev, who promptly let all his friends read it. Everybody has a wonderful time, especially when Makar Alekseevič tries to remonstrate with Ratazjaev:

I told him that he was a traitor. And Ratazjaev replied that I was a traitor myself, what with all those *conquêtes* of mine. He said: "Why, you've really kept it a secret! We never knew you were a Lovelace!" And now they keep calling me Lovelace and not by my own name! Do you hear my little angel, do you hear? They know it all now, they know all about it, and they know about you, my dear, and whatever you have, they know about it all! (I, 168).

Needless to say that Ratazjaev immediately proposes to immortalize the romance of Devuškin and Varen'ka in one of his "brilliant" satirical works. And there it is, the sentimental romance — reduced to a travesty! Meanwhile, Devuškin proceeds to "develop" his end of the travesty. He discovers that he has some rivals, various officers as well as civilians who are attracted by Varen'ka's beauty, defenselessness, and, alas, slightly tarnished reputation. He rises to her defense and promptly makes a fool of himself. He gets drunk and goes to see one of Varen'ka's admirers, an officer, "to tell him off". Here is how this adventure ends:

I don't remember what I said either. I only know that I said a good deal in my noble indignation. Anyway, that's where they turned me out, that's where they threw me down the stairs, or rather, it wasn't that they really threw me, no, they kind of shoved me out. (I, 152)

Had they at least *thrown* him downstairs! That would have made it more "honorable", for even gentlemen get thrown out of places on occasion, yielding

[6] W. Komarowitsch, *F. M. Dostojewski — Die Urgestalt der Brüder Karamasoff* (Munich: R. Piper & Co., 1928), pp. 177-181, points out a number of echoes from *Julie* in *Poor Folk*. Devuškin's report about some "disorders" involving him with a group of officers (see my next quote) is another such echo.

to superior forces. But never would a gentleman let himself be "shoved out" like a piece of furniture! Discouraged by this failure, Devuškin only wrings his hands when the next suitor, a substantial, middle-aged civilian, makes his appearance. "These scoundrelly libertines", he exclaims, "are intent upon doing me in; and they will, I swear, they will!" (I, 160).[7] In his excitement he vouches for the reliable efficiency with which his enemies are going to dispatch him, congratulating them on it in effect.

When Bykov arrives, it is the ridiculous ease with which he destroys Devuškin's dreams that makes a joke out of the whole "conflict". Bykov has heard a lot about Varen'ka's admirer, of course, and at first seems quite interested in learning more about him. But then Fedora points him out to Bykov, as he walks across the yard. Bykov only laughs and shows no further concern (I, 192). Varen'ka accepts Bykov's proposal with grateful, almost indecent alacrity. From here on the faithful Devuškin's role is limited to helping in the preparations for Varen'ka's wedding. Of course he volunteers for this role — the most cruel of the many little cruelties which make up this travesty of a great love.

Closely linked to the travesty is a kaleidoscopic sequence of literary parodies which accompany the action of the novel. Much as Werther and other sentimental heroes react to the literature of their days, so does Devuškin to the literature of his own time. Both he and the literature he reads are made quite ridiculous in the process.[8] Devuškin's very first letter contains a — rather gentle — parody of the still fashionable Romantic poetry *à la* Lermontov. One of its most popular images appears in the line "Why am I not a bird, a bird of prey aren't I!" which Devuškin "quotes". The prosaism "bird of prey" contributes just as much to the humorous effect as does the homely paraphrase of the whole poem.

The works of the versatile Ratazjaev from which Devuškin quotes several excerpts in his letter of June 26 stand for three different schools of Russian literature in the thirties and forties. The "historical" novel *Ermak and Zjulejka* represents the pseudohistorical style of, say, M. N. Zagoskin. The romantic adventure novel *Italian Passions* is apparently aimed at N. A. Polevoj more than

[7] Devuškin actually uses the word *lizobljudnik*, 'flunky' (lit. 'a lickplate'), but "libertine" must be what he means. The original meaning of "libertine" is, of course, "freedman, former slave", whence, apparently due to misunderstanding of *Acts*, VI: 9, "profligate" (Ernest Weekley, *An Etym. Dict. of Modern English*, [New York, 1921]); so also in French. Older French-Russian and English-Russian dictionaries invariably list the former meaning ("former slave"), which is very close to "flunky", first. I presume that Dostoevskij is making fun of some unfortunate translation of an eighteenth-century novel, in which "libertine" may have been translated by *lizobljudnik*.

[8] On the parodies, see Dmitrij Tschiżewskij und Johann Schröpfer, eds., *Russische literarische Parodien* (= Heidelberger slavische Texte herausgegeben von D. Tschiżewskij und Johann Schröpfer) (Wiesbaden: Otto Harrassowitz, 1957), pp. 33, 38-39.

at A. A. Bestužev-Marlinskij. The short "humorous" passage could have been written by Gogol' himself, as I will presently show. Devuškin's innocent admiration exposes the weaknesses of these literary trends as much as does the parody itself, which, incidentally, shows excellent style and good taste throughout: it exaggerates the traits of the originals by the merest trifle, just enough to make them ludicrous, yet without doing them an injustice.

Ermak, "a wild and terrible Cossack from the days of Ivan the Terrible", immediately produces a tirade in the purest style of a second rate Gothic novel of the 1830's. The point is that the anachronism does not in the least destroy Devuškin's illusion. What Makar Alekseevič finds so beautiful about Ratazjaev's *Italian Passions* is its "sentimental 'high style'" (to use Dostoevskij's own expression), saturated with trite poetic metaphors and unbelievably hyperbolic imagery, the ludicrous incongruities of which he naturally does not notice. The point of the third parody is that Devuškin finds this inane banter so funny. Naturally there is nothing funny about it (except, perhaps, its stupidity), much less anything that might be justly called "a little fanciful and perhaps rather too playful", as Devuškin puts it. Obviously the thing is a parody of Gogol's *Story of How Ivan Ivanovič Quarrelled with Ivan Nikiforovič*. Ratazjaev's Ivan Prokof'evič Želtopuz is an echo of one Anton Prokof'evič Pupopuz who appears in Gogol's story. Ivan Prokof'evič immediately goes on to bite the leg of one Prokofij Ivanovič. In Gogol's story it is one Agafija Fedoseevna who bites off somebody's ear. Prokofij Ivanovič, victim of this assault, is said to be very fond of turnip with honey. In Gogol's story the heads of the two antagonists resemble a turnip. The style of the passage neatly parodies Gogol's verbal "clowning", his tendency to present inanities as gems of humor, his sham folksiness.

Gogol's *Overcoat* is the target of a massive frontal attack by Makar Alekseevič himself. The entire long letter of July 8 is devoted to it, as well as a number of allusions elsewhere, always in the same angry tone. The letter of July 8 is of pivotal importance. With that letter, i.e., with the reading of *The Overcoat*, Devuškin's troubles really start. From here on there begin to appear, one by one, Dostoevskij's several proofs to the effect that Devuškin *is not* Bašmačkin. But in the letter itself Makar Alekseevič admits, implicitly and explicitly, that he is indeed Bašmačkin, point for point, down to walking on tiptoes to save the heels of his shoes! His vociferous protestations to the effect that Gogol's story is "subversive" and "fantastic", but also "nothing but another, trifling example of vulgar everyday life" (of course he does not see that he is contradicting himself), as well as his suggestions as to how the plot of *The Overcoat* ought to be changed, make the label "Akakij Akak'evič" stick so much more firmly.

Contrary to *The Overcoat*, Puškin's *Stationmaster* meets with Devuškin's enthusiastic approval. Here, the joke is entirely on Makar Alekseevič, as

Dostoevskij lets him make some "improvements" in Puškin's text. In Puškin's tale the stationmaster wipes away his tears with the skirt of his coat; in Devuškin's paraphrase he does it with the "*dirty* skirt of his coat". The picturesque naturalistic details of Vyrin's drunkenness also belong to Devuškin, not to Puškin. Altogether, Devuškin's version of the tale is a fine example of sentimental "low" style. The Natural School of the 1840's is of course the butt of this parody.

Miscellaneous witticisms of a literary nature occur throughout the novel. Thus, for instance, in his letter of April 8 (I, 82) Devuškin mentions Homer in one breath with — Baron Brambeus, of all writers. His eulogy of the organ grinder in his letter of September 5 (I, 178) is a "dig" at D. V. Grigorovič's physiological sketch *The Organ Grinders of St. Petersburg* (1844).

There are a few funny situations in the novel: Devuškin peeping through the keyhole as Aksentij Osipovič and Pëtr Petrovič complete their little *affaire d'honneur* to the satisfaction of both parties; Devuškin's visit to the money lender who "has got no money" (for Devuškin); Devuškin "exposed" as a "Lovelace"; the button scene in letter of September 9. But mostly the humor is verbal, based on Devuškin's unsuspecting naïveté and imperturbable bonhommie in the face of a puzzlingly complex and hopelessly indifferent world.

Alas, it is usually on the receiving end of the joke that we find our hero, with the world around him scoring its points, often painfully enough. The humor is then strictly one of *Schadenfreude*. We see Devuškin taken advantage of — and being grateful for it, for instance, when he copies Ratazjaev's works in exchange for the great man's "benevolence". Here is how he sees it:

I am doing some copying for him. But don't think, Varen'ka, that there is anything wrong with that, or that he is benevolent to me only because I am copying his works for him. Don't you believe such idle gossip, my dear, don't believe any such vicious gossip. No, I am doing this of myself, of my own accord, and for his pleasure. And as to his being benevolent to me, well, that's what he does for my pleasure. (I, 130).

And so Devuškin trades his own hard work for the benevolence of its recipient, that is, "for love". Then again, we see him making inadvertent revelations, injurious to himself or to Varen'ka, never knowing that he has once more talked too much. Thus, in his very first letter, Makar Alekseevič tries to minimize in every possible way the inconvenience of his new lodgings (he has moved there only to be near Varen'ka!), but in the process blurts out "Well, what a hole have I got into!" When he describes his "room", it turns out that he is living — in the kitchen. He corrects himself quickly, adding "or rather, to put it more accurately, I live ..." — and finishes by explaining how he lives in the kitchen after all (behind a thin partition), and not "near it". And so, time and again, the talkative Makar Alekseevič reveals unwittingly what he was really trying to keep a secret: that he has sold his uniform, that he owes the landlady some rent, that he *is* doing some gratis work for the worthless scribbler Ratazjaev and, last but not least, that he is desperately in love with Varen'ka.

Sometimes Makar Alekseevič's naïve parroting of everything he may have heard from others or that may have come to his mind is bound to embarrass Varen'ka. I have already quoted an example to such effect. In another instance, Varen'ka suddenly finds herself "disrobing before everybody" — as a figure of speech, of course (I, 154). In still another, she must hear that she has been called "certain indecent names" by the landlady and that, altogether, everybody knows her as "the babe who got tied up with that old devil" (Devuškin being the "old devil").[9] And all this poor Varen'ka must hear from her faithful admirer, of all people. The joke is on her, delicate, dainty, sentimental Varen'ka. The *Schadenfreude* is only the juicier for it.

Devuškin is the typical "loser". We know, everybody knows, he himself knows that whatever he will try to do will be a failure. He makes a laughing-stock of himself by even trying. What chance does he have to get a loan from Markov the money-lender? Without a security, his uniform almost in rags, his shoes full of holes, and his generally wretched appearance suggesting chronic impecuniousness? What chance does he have with his protests against a bunch of young officers, at the flat of one of them? Still, he goes there "in his noble indignation". What else could happen but that they would throw him out, with ignominy and without much ado? Or don't we know that something positively ludicrous will happen when Devuškin is dragged before His Excellency, even though he will surely be unable to utter a single word during the whole interview? And sure enough, he loses a button, dashes all over the place trying to recover it, then stands there, smiling like an imbecile and trying to fit the button to the torn threads. And he never says a single word, even though a lot of things happen before he finally leaves the office. As a consistent and persistent loser, as a perennial failure, Devuškin generates a laughter of *Schadenfreude* — which is the laughter of secure superiority — comparable to that which rewards the "sad" circus clown who goes through a routine of ludicrous failures and misfortunes.

Next to Devuškin the loser, it is Devuškin the simpleton who makes us laugh. His view of the sophisticated and institutionalized world in which he lives is, in spite — or because — of his good common sense and homely wisdom, often absurdly naïve. "What a capital thing literature is, Varen'ka", he exclaims in his letter of June 26, "truly a capital thing. I heard that the day before yesterday. A profound thing. Strengthening men's hearts and many other things that are said about it in that book of theirs" (I, 130). Well, one can't blame Makar

[9] The original reads *chërt s mladencem svjazalis'* (I, 155).

Alekseevič for having a naïve view of literature — many fine people never do find out what a capital thing it is. But here is the titular councilor's view of the world in which he ought to be at home:

I have been in the service for nearly thirty years; my record is irreproachable, my conduct has been marked by temperance, nor has any irregularity been ever put down to me. As a citizen I consider myself in my own mind as having my faults, but my virtues too. I am respected by my superiors, and His Excellency himself is satisfied with me; true, he has hitherto not shown me any special marks of his favor, yet I know that he is satisfied. My hair is now grey. Still, I am not aware of any major sin that I might be guilty of. Naturally, no one is free of minor failings. We're all sinners, even you are a sinner, my dear. But no serious offense or gross violation of discipline has ever been recorded against me, I mean, such as anything against regulations or disturbance of public tranquillity, no, nothing of that kind. (I, 144-145).

In a word, he has understood nothing at all of the scheming, conniving, hypocritical, thoroughly predatory society around him, where rules and regulations are made so that sheep, like Devuškin, can be clipped with impunity, where Devuškin's qualities of plodding assiduousness and humble obedience take a man exactly nowhere. After thirty years of faithful service, His Excellency hasn't even heard of Devuškin, much less noticed him. When Makar Alekseevič hears one of his superiors say that to know how to make money is the most important civic virtue, this is news to him! (I, 125). At almost fifty years of age he has retained the innocence of a babe.

Yet this same good-natured simplicity of Devuškin's sometimes scores against the very sophistication and complexity of the imperial bureaucracy. The following discourse on "human relations" requires no comment:

Why, inasmuch as there are various grades in the Table of Ranks there also should be reserved for each grade its very own, appropriate sort of tonguelashing to be administerred according to rank, so that, by course of nature, the tone of any given tonguelashing would be accordingly graded by rank — such is the order of things! Why, the whole world rests on nothing but that, my dear, namely on everybody's trying to impress the other, and on everybody's administering tonguelashings to someone else. (Letter of July 8, I, 146).

It's just as simple as that! And so, without knowing it, Devuškin does "get even" with the imperial bureaucracy. He also "gets even" with Ratazjaev and literature. When Varen'ka suggests that Ratazjaev may not be such a great writer after all (she even doubts if he has ever printed a line), Devuškin produces an apology of sorts, in which he tries to say that while there may be, of course, writers who are better than Ratazjaev, even much better, Ratazjaev is still very good, so that while these writers write very well indeed, Ratazjaev writes well also. "Well, anyway, he just keeps scribbling away", Makar Alekseevič concludes his harangue, "and it's certainly all to the good that he does keep scribbling" (I, 138). Devuškin's unintentional (is it really unintentional?) pun involving the verb *popisyvat'* which may mean "to write a little, to write occasionally", but also "to scribble", is untranslatable.

Sometimes there is wisdom in Devuškin's naïveté. Sometimes the very goodness of his heart scores humorous effects. Take, for instance, Devuškin in the role of benefactor, with Gorškov (formerly an official of some consequence, for how else could he have become involved in a major scandal about some government contracts?) grovelling before him in an effort to obtain the loan of ten kopecks! (I, 182-184). Gorškov's stilted rhetoric and extravagant politeness are as ludicrous in relation to the "sum" in question as is Devuškin's condescending, dignified benevolence. Or, what would His Excellency say if he knew that his titular councilor refers to him as "the poor dear" — in connection with a scene where he "deigned to be stern and was very angry with Emel'jan Ivanovič and shouted so much he finally was quite exhausted, the poor dear" (I, 203).

Not always is Devuškin's humor entirely involuntary. He can be simply playful, in a goodnatured way, as for instance when he imagines himself walking down Nevskij Prospect as "Devuškin, composer of literature, and bard", then adds as an afterthought: "But then, what about those torn shoes of mine? ... What if all those people learned that Devuškin, *littérateur*, has patches all over his shoes? What would her ladyship the *comtesse-duchesse* say if she heard that?" (I, 134). He can also play the buffoon (this usually happens when he has had one too many), as in that letter of August 19, where he proposes to forget about wearing shoes altogether — "for didn't the sages of antiquity walk around barefooted?" (I, 171), or reports that his friend Emelja was trying "to pawn a certain object which really can't be pawned, for there is no such pledge as that" (I, 172) — it seems that his *sine quibus non* are meant.

Finally, there are those rebellious letters of August 1 and September 5 in which the normally timid, humble Makar Alekseevič displays the angry irony of impotent plebeian resentment against the *hubris* of *hoi aristoi*:

Now, why does a poor man know all this, why does he think of it like this? Why? — from experience, of course! Why, because he knows, for instance, that right next door there lives this gentleman who, at this very moment, on his way to a restaurant is saying to himself: "I wonder what this beggar of a *činovnik* is going to eat to-day? As for myself, I shall eat *sauté-papillote*. Maybe he will be eating porridge without butter." Now, what business is it of his that I am going to eat my porridge without butter? (I, 154)

Makar Alekseevič is trying to be sarcastic! Alas, he does not succeed. The rich man will be eating his "fried paperwrappers" (*sauté-papillote* is a hybrid of two culinary terms) undisturbed. The joke is on Makar Alekseevič and on the poor man who will have to eat his porridge in the unenviable knowledge that the rich man is gloating over his plight. The same pattern of irony that "backfires" appears in the other displays of Devuškin's rebellion.[10]

As to Ermilov's contention that the naïve conformism of our hero is systematically set against the ugly reality of life under Nicholas I, thus forming a definite pattern of what Ermilov calls "author's irony", I feel that my own analysis does not confirm this thesis.[11] The instances of ironic insinuation apparent in some of the above examples and in a number of others do not, I think, follow enough of a specific pattern to convey a message of protest against the existing social and political order. Much rather, I would say, could there be abstracted, from *Poor Folk*, a pattern of "protest" against the *inherent* inequality of man.[12] Dostoevskij's humor in *Poor Folk* is not satirical, except perhaps in the few passages of direct literary parody. It is rather a humor which is derived from the eternal conflict between the simple soul of a plain man and the complex apparatus of a soulless, institutionalized society run by "clever" people. There is enough of Sancho Panza in Makar Alekseevič Devuškin. Like Sancho, Makar Alekseevič, usually funny and often ridiculous himself, manages to make the world around him look quite ridiculous also. If the humor in *Poor Folk* is, in part, that of satire, it is of satire which deals with eternal human problems much more than it does with the problems and conflicts of Russia in the 1840's.

The humor in *The Double* is of a very different sort. Goljadkin has no sense of humor.[13] Even in his passive role as the butt of the Doppelgänger's and the narrator's jokes he behaves quite differently than Devuškin in a similar role. The latter would try, awkwardly, to parry the taunts aimed at him, thus increasing the comical impact of the "score", much as a boxer who keeps running into his opponent's punches. Goljadkin, too, "cooperates" with his tormentors, but in an entirely different way. His recurrent fits of impotent rage and mad rushes at a phanton enemy remind one of a groggy fighter who, blinded by his own blood, keeps flailing away at an elusive opponent. If Devuškin is a "sentimental" clown, a clown with a heart, Goljadkin is the image of a schizophrenically humorless, mechanically jerky, puppet-like "dead pan". Devuškin

[10] E.g., when Devuškin wants to give an example of a poor, but upright and honest man, who would not depend on charity, but prefers to work hard for his living and to be useful to society, he mentions — an organ grinder, of all people! (I, 178).

[11] V. V. Ermilov, *F. M. Dostoevskij* (Moscow: Gosizdat, 1956), pp. 55-56.

[12] This is how Apollon Grigor'ev seems to have seen it. See Apollon Grigor'ev, *Polnoe sobranie sočinenij i pisem* (Petrograd: P. P. Ivanov, 1918), p. 128.

[13] N. E. Osipov, "*Dvojnik. Peterburgskaja poèma* F. M. Dostoevskogo" (A. L. Bem, ed., *O Dostoevskom*, I [Prague: Ferd. Svoboda, 1929]), pp. 40-41, points out some instances of Goljadkin's "humor". However, Osipov (who is dealing with *The Double* from a psychiatric viewpoint) specifically defines these bits of "humor" as psychopathic symptoms.

makes us laugh by "finding" new and different ways to ask for more punishment. Goljadkin overwhelms us by the sheer quantity of punishment (in the form of taunts, irony, sarcasm, as well as "practical jokes") which he absorbs before finally collapsing.

First of all, the basic mode of the narrative in *The Double* is, in my opinion, definitely ironical.[14] The hero is pursued by the mocking echo of his own words and thoughts not only from his elusive Doppelgänger but also from an ever-present, methodically cruel narrator.[15] With imperturbable "bonhommie" and concerned "sympathy", the latter keeps presenting Goljadkin's hallucinations as perfectly normal experience, his disorganized ramblings as logical reasoning, and his frenzied dashes into the void as well planned, sensible activity. The "benevolent" approval of the narrator does not fail Goljadkin even when he is to commit the greatest folly of all, the "elopement" with Klara Olsuf'evna. When, for the last time, reason returns to Goljadkin and he is about to desist from his mad scheme the narrator reports, in the tone of profound satisfaction, that his hero did not waver long, "but having gotten as far as Semënovskij Bridge, judiciously decided to return" and go through with it after all (I, 367). As so often, he calls Goljadkin's madness "judicious".

Not infrequently, the narrator's mock bonhommie turns into outright derision, a circumstance which caused Mixajlovskij and others to refer to *The Double* as to the first product of Dostoevskij's "cruel talent".[16] When Goljadkin is ejected from the ball the narrator spares him no painful detail of this humiliating experience. He clothes the account of Goljadkin's ejection into a series of euphemisms which are, however, so transparent and suggestive that, rather than mitigating the hero's embarrassment, they actually add insult to injury. Toward the very end of the novel (I, 359, 369) there appears an "echo" of this scene — and the irony is still there! It is here, toward the end of the story, when Goljadkin is utterly helpless, exhausted, and miserable, that the narrator's continued irony begins to smack of cruelty. When Goljadkin, already driven to despair by his many misfortunes, decides to take recourse to an interview with His Excellency he is greeted — why, by the Doppelgänger, of course, who is given the epithet "Mr. Goljadkin's close acquaintance and friend" (I, 357).

[14] Here I disagree with V. V. Vinogradov, "K morfologii natural'nogo stilja: Opyt lingvističeskogo analiza Peterburgskoj poèmy *Dvojnik*", *Èvoljucija russkogo naturalizma: Gogol' i Dostoevskij* (Moscow: "Academia", 1929), p. 218, who believes that "a factual (*delovoj*) narrative style, with a shade of humor, which describes the most minute movements and emotions of the hero, as well as details of the setting, forms the basic foundation of the novel".

[15] I am following M. M. Baxtin, *Problemy tvorčestva Dostoevskogo* (Leningrad: Priboj, 1929), p. 152 (and *passim*), who sees *The Double* as a "dialogue" between the two Goljadkins and the mocking echo of the narrator.

[16] N. K. Mixajlovskij, "Žestokij talant", *F. M. Dostoevskij v russkoj kritike* (Moscow: Gosizdat, 1956), pp. 343-346.

A little later the narrator alludes to his hero's "proven firmness and resoluteness" (I, 360), and calls his confused blundering a "felicitous plan of action" (I, 367). In the following passage we see Goljadkin cowering behind a stack of firewood in Berendeev's back yard, waiting for Klara Olsuf'evna:

Of course, Spanish serenades or silken ladders were now out of question; but a secluded spot, snug and private, if not altogether warm, certainly was not. He felt greatly tempted, it might be mentioned in passing, by that very corner in the back staircase of Olsufij Ivanovič's flat where our hero had once, almost at the beginning of this veracious story, stood for more than two hours between a cupboard and an old screen, among all sorts of domestic and sundry trash, rubbish, and litter. (I, 360-361)

The mention of "Spanish serenades" and "silken ladders" (an allusion to Puškin's poem "I'm here, Inezilla"?), while ironic, is not cruel, for Goljadkin has given up that fancy anyway. But the litotes "not altogether warm" is, for it is beastly cold out there — in the yard as well as in the back staircase. Note the insidious way in which the narrator feigns full solidarity with Goljadkin: "certainly" (*nužno že bylo*) belongs to the narrator! The conventional ironic reference to the veracity of the story (we find it often in *Don Quixote*, for instance) is found repeatedly in the novel. Finally, it is with undisguised glee that the narrator recalls a very painful past experience of Mr. Goljadkin's. Note how he dwells upon the unpleasant details of that episode!

The striking use of motor images, usually expressed by verbs of motion, in developing the plot and in drawing the characters of the novel was first pointed out, and described in detail, by Vinogradov.[17] These elements are so numerous, so elaborately organized, and so continually present throughout the novel that they form, one might say, a "second plot", which moves along on a strictly visual plane of motor images and is superimposed upon the basic, stream-of-consciousness type narrative. This "second plot", if abstracted from the novel, would form something along the lines of a grotesque, superdimensional puppet play, anticipating in many ways the mood and the technique of a Charlie Chaplin movie. The basic "motor pattern" is, for Goljadkin Senior, that of a jerky zigzagging around, with brief, frantic, headlong dashes alternating with moments of paralyzed immobility:

Having thus resolved the matter of his present condition, Mr. Goljadkin quickly moved forward as though someone had released a spring in him; in two steps he found himself in the refreshment room, flung off his overcoat, took off his hat, hurriedly stuffed these things into a corner ... then he moved on to the tea-room, from the tea-room he darted into still another room, slipped almost unnoticed between groups of cardplayers engrossed by their game ... then ... here Mr. Goljadkin forgot everything that was going on about him, and burst into the ball room like a bolt from the blue. (I, 242)

[17] V. V. Vinogradov, *op. cit.*, pp. 279ff.

Suggestions to the effect that Goljadkin moves like a puppet ("as though someone had released a spring in him") are frequent throughout the novel, as are examples of rapid, "cinematic" motion: Goljadkin's legs are very short, so that, to get from one room to the other "in two steps", he must be moving like a jack-in-the-box. The jerky, jumpy quality of Goljadkin's movements is also brought out by the jerky syntax of the narrative: anaphoric gradation, repetition of "then", and ellipsis, all of which appear even more clearly in the original. When visualizing Goljadkin's movements one must not forget that he is middle-aged, very short, and quite stout — not the normally agile type. The more surprising, and comical, to see him in sudden, rapid motion! When Goljadkin and his Doppelgänger arrive at Goljadkin's house almost at the same time, "the gate opened, the stranger stooped, darted in, and vanished; almost at the same instant, Mr. Goljadkin reached the spot and, like an arrow, flew through the gate" (I, 256). The stranger, being a ghost, might well "dart in and vanish", but how can the shortlegged, paunchy Goljadkin "fly like an arrow"? Anyway, throughout the narrative, we see Goljadkin "in full pursuit of his enemy", "jumping off the porch in one leap", "flying along". And here, for a final example, is the following, very funny scene:

> There was some noise ... Mr. Goljadkin pulled his head between his shoulders and leaped behind the stove. Some one came down the stairs and went out into the street. "Who could it be, leaving the department at this time?" our hero thought to himself. A moment later someone else's footsteps were heard ... Here, Mr. Goljadkin could not restrain his curiosity and, from behind his rampart, poked out the tiniest, the very tiniest tip of his nose — poke out he did, but also withdraw it in a flash, as though somebody had pricked it with a pin. (I, 324)

Again, a scene which is all motion. Its humor is of the puppet theatre variety. The comic effect of a nose popping out from behind the stove and disappearing with lightning speed would, of course, be perfect if produced by the outsized nose of a puppet.

In contrast to Goljadkin Senior who alternately moves at breakneck speed, then freezes in his tracks, Goljadkin Junior is always "on the move". He is characterized by ceaseless activity, quick and adroit motion. He is "going places", literally and figuratively, with nary a moment of rest. He is labelled by verbs of motion such as *vykidyvat' nožkami* ("kick up one's legs"), *v'junit'* (dart around"), *ljagnut' nožkoi* ("to kick"), *nyrnut'* ("to dive"), *semenit'* (to mince"), *uliznut'* ("to slip away"), *šmygnut'* ("to dart"), *julit'* ("to bustle about"), and *jurknut'* ("to whisk"). Most of these verbs have a funny connotation *per se*. All of them occur repeatedly, and with different prefixes which still increase the effect. Thus, Goljadkin Junior can be a very entertaining character, if one can only visualize his funny antics. Of course there is also something uncanny about that dexterity of his, about that smooth way in which he keeps appearing and disappearing, about his extraordinary pep and hustle. Obviously the pattern of his movements is that of a petty demon — but a discussion of this point would take me beyond the scope of this paper.

To summarize the above: the grotesque puppet theatre of the "motor plot" in *The Double* has a great comic potential; if visualized, the antics of the two Goljadkins can be as funny — and as gruesome, of course — as only a puppet theatre can be; its humor is that which seems to be inherent in any mechanization of a human (or of any living) function.[18]

Goljadkin is, like Devuškin, a poor devil who is running the gantlet of a whole row of misfortunes and humiliations — to amuse the reader. Here, *Schadenfreude* again enters the picture, the difference between *Poor Folk* and *The Double* being that in the latter it is usually the narrator who does the "rubbing in", while Devuškin had had to do it himself. And yet, scenes in which Goljadkin, too, plays the involuntary buffoon are not rare. Take, for instance, that long internal monologue in chap. ix (I, 296-297) where Goljadkin is trying very hard to minimize the unpleasantness involved in having a Doppelgänger and is in fact trying to find advantages in it. The comicalness of Goljadkin "admiring" his own image as one half of a pair of identical twins is reminiscent of Puškin's gleeful image of the cuckolded man throwing furtive glances at his own horns in a mirror yet refusing, to the last moment, to admit the truth to himself (*Eugene Onegin*, iv, 33), or of Amphitryon being persuaded that to have been cuckolded by Jove is really a great distinction. No matter how hard the poor man tries to convince himself that having a Doppelgänger is perfectly natural and quite harmless, consistent with divine providence and, in fact, rather "touching", it remains a most unpleasant business — even if the higher-ups should approve of it. He simply can't work up any real enthusiasm for his new condition, and the end of this long "pep talk" to himself finds him as confused and unhappy as ever. In fact, deliberate discussion of the painful details of living one's life as one half of a pair of Siamese twins (what are people at the office going to say? and what about visitors to the department? higher authorities?) makes things appear worse, not better. In a word, Goljadkin presents as funny a picture as does anybody who is trying hard to extricate himself from an unpleasant predicament but instead gets entangled more and more.

Thanks to the narrator's irony some of the "adventures" of Mr. Goljadkin which, from the latter's viewpoint, are far from funny take on a decidedly farcical character. To Jakov Petrovič, his shameful failure at the ball is nothing

[18] "Les attitudes, gestes et mouvements du corps humain sont risibles dans l'exacte mesure où ce corps nous fait penser à une simple mécanique", says Henri Bergson in *Le rire*: *Essai sur la signification du comique* (17th ed., Paris: Félix Alcan, 1918), p. 30.

short of "catastrophe", and we know it (from the stream-of-consciousness account of the event, which presents his viewpoint). The narrator's description of some moments of the same episode, on the other hand, presents the funny side of Goljadkin's adventure: the young men who, it seems, have a hand in his ejection from the ballroom may well have seen it that way! A scene where somebody gets thrown out of a place can be great fun, or most unpleasant, depending upon one's own role in it. Dostoevskij lets his readers see it from both sides. Those to whom scenes such as that just mentioned seem too cruel to be really funny will easily find some which feature humor in a lighter vein. Goljadkin's encounters with his social inferiors — Petruša, Gerasimyč, the menials at the office — are pure comedy with veritable fireworks of funny "gags". Take the scene (I, 306-307) where Goljadkin is trying to learn from Petruša, his man, whether or not he has taken care of an errand from which he has just returned reeling drunk. Goljadkin desperately tries to shake him awake and get the facts from him. Petruša's answers, given in a drunken stammer, are hopelessly off their mark. With the assistance of Goljadkin's absurd "leading" questions he comes up with one "punch line" after the other. At the very outset, Petruša's stammering and inability to come up with a complete phrase convict him of perjury: he's had considerably more than "a drop", as charged by his master and denied, in juratory terms, by himself. As to Goljadkin's reaction ("forget it, it's very good that you are drunk!"), it probably has nothing to do with Goljadkin's mental condition: anybody trying to get a drunken person to his senses could have uttered it in desperation. Dostoevskij cannot resist the temptation to introduce such bits of verbal humor into his dialogue on occasion. The oxymoron "you do remember everything, don't you, only you have forgotten" is another example to this effect. On stage this, and other similar lines would, no doubt, draw laughter from the audience. A few other comical effects in this scene rest on the circumstance that Petruša keeps reacting not to the meaning of his master's questions, but to the tone in which the question has been asked. When he denies having been at Vaxrameev's, the tone of Goljadkin's question must have indicated that he should *not* have been there. A moment later, when Petruša realizes that his master wants an answer in the affirmative he promptly comes up with it. When asked about the address of the "other" Goljadkin, he gathers that he is expected to say "yes, I've got it", and does so — although he did not get any address at all. When Goljadkin insists on getting the address Petruša, who is now afraid of him, blurts out the only address of a "Goljadkin, titular councilor" that he knows: his master's. This is, of course, *the* punch line of the whole scene. The rest of what Petruša has to say about his visit with Vaxrameev is plain fiction. Vaxrameev has said nothing of the kind and, least of all, that Petruša is "a nice fellow". On the contrary, in his letter he warns Goljadkin of

Petruša! The whole scene is undeniably funny, yet without a trace of "cruelty". Of course Goljadkin is the "straight man" just as everywhere else in the novel, but he is neither abused nor taken advantage of.

In the scene between Goljadkin and Ostaf'ev, the drunken scribe, we witness another "communication breakdown" verging toward the farcical. For two pages and a half (I, 319-322), Goljadkin tries to pry some information about what has been going on about the office from the stupid and rather drunk Ostaf'ev and keeps getting such answers as "yes, your honor, I wish you good health, your honor", and "no, your honor, nothing so far, your honor, nothing at all". Every time he gives Ostaf'ev a ten-kopeck piece to loosen his tongue all he gets is a soldierly "I'm feelingly grateful for your honor's kindness", but no news. Finally, when Goljadkin is ready to give up, Ostaf'ev, quite inadvertently, blurts out: "Yes, your honor, and Ivan Semënovič took your place at the office to-day". So there had been some news after all. By the way, Ostaf'ev has a "label": being habitually drunk during office hours, he is greatly concerned that his superiors should not smell the liquor on his breath. During the interview with Goljadkin mention is made no less than four times of Ostaf'ev's covering his mouth with his hand. The reason for this mannerism is not stated — a finesse typical of Dostoevskij's craftsmanship. He lets the reader experience the pleasure of detecting this little trait. Altogether, this scene, too, would surely draw laughter if presented on stage. And again, Goljadkin is a "straight man" without, however, being pathetic.

More than in the case of *Poor Folk*, there is a question as to whether the wealth of humor found in *The Double* fits into the scheme of the novel. Elements of comedy, farce even; a grotesque puppet theatre, accompanied by an ironic narrative; the mocking echo of Goljadkin's stream-of-consciousness in the taunts of his Doppelgänger — all this cannot fail to bring out a ridiculous quality about Goljadkin's madness. Goljadkin's folly is ridiculous in its wretched pettiness, its total lack of spirituality, in its shallowness of feeling. It is, in a way, a travesty of madness. This means that the humor of *The Double* is, in a broad sense, satirical, negative. *The Double* is a grim satire on what Dostoevskij himself was later to call "a social type of tremendous importance".[19] The philosophical implications of this satire are beyond the scope of this paper.[20] Only one of its aspects may be mentioned here, being rather clear and unequivocal: Goljadkin is, among other things, a hypocrite who is exposed by the indiscretions of the Doppelgänger, and occasional insinuations of the narrator.

[19] In a letter to his brother Mixail, dated October 1, 1859 (F. M. Dostoevskij, *Pis'ma*, I, 257).
[20] They have been treated by Dragutin Prohaska, D. I. Čiževskij, A. L. Bem, more recently by Viktor Šklovskij and Konrad Onasch, to name but a few authors.

He "protests too much" and gradually manages to convince the reader not of his innocence, but of his guilt.

In *Gospodin Proxarčin* there seems to be no profound reason why the narrator should be making fun of his characters. Yet he does it all the time and apparently enjoys himself doing it. For the most part it is again pure *Schadenfreude* that he experiences, and the sympathetic reader with him. He seems to be having as much fun following the progress of the hoax played upon Proxarčin as he does reporting the well-deserved rebuffs dished out by the old man to the likes of Zinovij Prokof'evič and Mark Ivanovič. Every single character in the story, except perhaps Jaroslav Il'ič the police officer, gets the same ironic treatment. No mercy is shown to the inexperience of youth, or to the debility of advanced age. Nor is the fair sex (in the person of Ustin'ja Fëdorovna) spared.

Zimovejkin is not a character who would deserve tactful treatment, nor does he get it. Every misfortune of the little crook is for the narrator an occasion for unconcealed glee. His suggestive rather than charitable euphemisms and leisurely dwelling upon the unsavory details of the sycophant's many scandalous deeds, all in the disguise of a disinterested yet humane biographer's benevolent objectivity, are sarcastic to the core. This narrator has as much charity, or even objectivity, in store for Zimovejkin and company as does Dickens for his villains. The basic figures of speech are here litotes, euphemism, mock charity. Zimovejkin is "ingloriously banished" (*s besčestiem izgnan*, I, 400) rather than "thrown out", he "deprives" (*lišil, ibid.*) his friend Proxarčin of a pair of breeches rather than "steals" them, and appears as the latter's "seductor" (*obol'stitel', ibid.*) when they go out and get drunk together. When Zimovejkin makes his final appearance we are told that "one could see that Zimovejkin had spent the night in waking, probably engaged in some important labors" (I, 409). The sequence reveals that Zimovejkin's "important labors" consisted entirely in drinking vodka. In this instance, Zimovejkin has been called upon to "straighten out" his friend Proxarčin. He immediately addresses the sick man in a "brief but forceful speech" — another sarcasm, for "drunken babble" would have been more appropriate. And on it goes, in the same merry fashion, until Zimovejkin and his associate Remnëv are caught trying to rob the dying Proxarčin. The last reference to them is another mocking euphemism as we hear, briefly and literally in parenthesis, that the police arrived on the scene and, in passing, "placed Remnëv and Zimovejkin in care of the proper parties". (I, 417).

Ustin'ja Fëdorovna's roomers are all laughed at in the most uninhibited, cheerful way although, truthfully, they are not really a group that would ordinarily raise one's spirits — a bunch of unkempt, brutish, drunken minor clerks, or worse. There is, for instance, Ustin'ja's late "favorite" who "might have lived for many years as the most faithful of her myrmidons and hangers-on, had he not finally drunk himself to an early grave in the most melancholy fashion" (I, 391). Here, the pun involved in *prispešnik i prižival'ščik* cannot be rendered in English. But even so, the easy jocular way in which this rather sad development is treated is quite remarkable. Then there are Mark Ivanovič, a stupid and loquacious *raisonneur*, Zinovij Prokof'evič, the "wise guy", who talks a lot about his future career in high society, and other assorted clerks with such funny names as Okeanov, Prepolovenko, and Oplevaniev. The last mentioned name, by the way, is used for a pun: "*Oplevaniev pljunul*" (I, 411). All of them are mercilessly made fun of, mostly in that they, a bunch of scribes and menials, are ostensibly treated as gentlemen of honor, culture, and refined wit.

"They liked", we are told, "to get the gang together to enjoy, as they liked to put it, life's scintillating moments" (I, 391). Again, we are dealing with a pun which gets lost in the English translation: *šipučie mgnovenija žizni*, "life's scintillating moments", carries a strong association with *šipučee*, "champagne", suggesting that they are, in fact, drinking that noble vintage at their gatherings. They are drinking vodka, of course.

Then, the crude hoax which they are playing on the ignorant old man is presented throughout as an exhibit of fine ironic wit. When they realize that they have driven their victim to despair, grave illness, and possible death, they make things worse, rather than better, by some truly asinine attempts to "cure" him of his *idée fixe*. The narrator rewards them with words of the highest praise for their "strenuous labors," "generosity", and "compassion". But most of all, the narrator gloats over the grief of these worthy gentlemen at the loss of all of Proxarčin's money to the police. Perhaps he enjoys the lamentations of Ustin'ja Fëdorovna, the landlady, even more. All of it, naturally, in the guise of the dispassionate historian's dutiful recording of the facts. Note the loving description of all that money (in the tender diminutive: *denežki!*), as it is being counted by the police officer, gathered into a huge pile, and taken away, with the roomers standing around, in sad resignation. Note also the laconic mention of the fact that Ustin'ja Fëdorovna was advised as to where she ought to submit her claim for rent outstanding (I, 420). Her prolonged wails show that she does not accept the loss with the stoicism of her clients.

The hero, Semën Ivanovič Proxarčin, does not get off easily either. The narrator, but also the other roomers, display a good deal of ingenuity in extracting laughs from what little the taciturn, gruff old man does and says. Thus, Mark Ivanovič, in one of his exhibits of "beautiful, flowery oratory", characterizes Proxarčin as "a nice and quiet person, although not a man of the world, loyal, not a flatterer, who has of course his faults but who, if ever the time should come for him to suffer, would suffer from nothing but his own lack of imagination" (I, 392). The narrator immediately parodies Mark Ivanovič's silly statement by pointing out, on his own part, that Proxarčin, "being thus deprived of any imagination of his own, could not easily impress one with his physical presence or polished manners either, yet managed to gain favor even without these qualities" (I, 392).

The story takes its course. Proxarčin is made to believe that he may be losing his job. Not so bad, in view of his ample savings, but for Zinovij Prokof'evič's "hunch", half facetious to be sure, about some money being hidden in Proxarčin's chest. This sudden threat really touches the old man to the quick: now his whole existence is in jeopardy. His worries, his clumsy attempts at forestalling the threatening danger, his desperate sorties against those whom he falsely believes to be his principal enemies — all this is to Proxarčin's "friends", including the narrator, nothing but a capital joke. Even Proxarčin's death, with all its macabre details, seems to amuse the narrator, inspiring him to a number of scurrilous jests. When the dead man is laid out in state, "dressed in his best although, to be sure, also his only suit of clothes", this is for the narrator merely another occasion for all kinds of puns and "wisecracks".

The narrator of *Gospodin Proxarčin* manages to draw an assortment of humorous effects even from "dead" verbal material by ingenious catachresis of the simplest, most prosaic phrases and expressions. Just one example:

> Ustin'ja Fëdorovna herself had seen, with her very own eyes, and with the aid of the decrepitude of Mr. Proxarčin's bedscreens, that the poor dear at times would have nothing to cover the nakedness of his body. (I, 393).

It takes ingenuity to find an expression such as "with the aid of the decrepitude of the bedscreens" instead of a simple "through a hole in the old and torn bedscreens". My translation "nakedness of his body" corresponds to *beloe tel'ce*, "white body" (with "body" in a hypocoristic diminutive), a stock expression of Russian folk poetry, which is grotesquely out of place here, considering the description of the owner of the "white body". This and numerous other examples of affected diction show that the narrator treats language much as he treats his characters: as material on which to exercise his ironic wit. I feel that here the young Dostoevskij has gone farther than even Gogol in his Petersburg Tales. In *Gospodin Proxarčin*, the irreverent treatment accorded to the Russian language by the narrator borders on abuse, on sacrilege, if one considers language a gift of God.[21]

[21] Clemens Brentano's warning to E. T. A. Hoffmann could very well be applied to Gogol and to the young Dostoevskij: "Über das liebe Brod liesse sich etwas ungemein Frommes und Wirkliches schreiben, über das Wort auch. Die Mythe jener Hungersnoth, wenn man mit

In short, the humor of *Gospodin Proxarčin* is that of nihilism, of absolute disregard for all values — but without any lofty philosophical justification. So, "cynical" is perhaps the correct term for it. Somehow this squares with what seems to be the idea underlying this story, the idea of man's absolute aloneness in a totally unpredictable world, an idea which the stupid, brutish Proxarčin, of all people, realizes and lives with iron consistency. Proxarčin then represents the tragic, the narrator the comic aspect of the same *Weltanschauung*.

While *The Landlady* is a thoroughly humorless work (even irony, normally Dostoevskij's trademark, is absent, except here and there in connection with Jaroslav Il'ič, a minor figure), the sad story *A Faint Heart* does not lack the quality of humor. In the earlier part of its simple, unpretentious narrative a good-natured sort of humor does in fact prevail. Obviously the narrator is very fond of both Vasja and his friend Arkadij Ivanovič. And so he makes fun of them, and their touching friendship, in a gentle, friendly way.

Arkadij Ivanovič is a jolly, fun-loving young bear of a man, always bent on practical jokes at the expense of his fragile, delicate roommate. When Vasja is forced to confess the secret of his young love while thrashing around helplessly in the lap of his strong-armed friend (a very funny scene!) the narrator cannot quite approve of such childish horseplay, but makes it clear that Arkadij Ivanovič really means well. When the secret is finally out, both friends present a very comical picture as they are staring at each other open-mouthed, unable to say a word for too much joyous, happy excitement. And the "strong man" of a few seconds earlier is now ready to cry. The fun and good humor reach a climax when Arkadij and Vasja enter Mme. Leroux's fashion shop to buy a present for Liza. Here is Vasja as he discovers the cute little bonnet which he is going to buy:

"Arkaša!" he said in an undertone, glancing in a casual way at all the grand and gorgeous things displayed upon little wooden stands on the huge counter of the fashion shop. "Wonders! What is this? And this? Look at that sweet thing over there, for instance; do you see it?" he whispered pointing at a very pretty little bonnet, but not at all the one he really wanted to buy. For he had already made his choice from afar, and was now drinking in the sight of *the* bonnet, the famous, the only real bonnet, displayed on the other end of the counter. The way he stared at it one would have thought that someone was about to steal it, or that the bonnet itself was getting ready to fly away from its stand, just so Vasja couldn't get hold of it. (I, 527)

Brodkugeln schiesst, werden Sie gewiss bei dem Wort auch finden. Die witzigen, gaukelnden, sogenannten Humoristen treten immer in der Literatur ein vor der Hungersnoth. Es ist das Henkersmahl, der letzte Schmaus des verlorenen Sohnes." (Quoted after Hans von Müller, ed., *E. T. A. Hoffmann im persönlichen und brieflichen Verkehr*, II, Briefe 1814-1822 [Berlin, 1912], 256).

And all that delight about a little eight-ruble bonnet! The humorous absurdity of the last image is suggestive of Vasja's naïve delight. No doubt he is very funny in his role of fashion expert. When the purchase is finally made, Vasja is so beside himself that he actually gives Mme. Leroux a kiss. The narrator comments that "it positively took all the great poise of that lady to preserve her dignity before this young rake" (I, 528). The use of the word *povesa*, "rake", for Vasja, a shy and innocent young man, is both humorous and affectionate. The narrator is glad to concede to Vasja and his friend these few moments of uninhibited happiness. Almost in the same breath, Vasja calls himself a "viveur", and the narrator, with a smile, lets it go.

The tone of light-hearted banter and good-natured "ribbing" prevails until the story begins to take a tragic turn. As soon as the reader has been told that Vasja is not really as happy as may have appeared, that a heavy worry is bothering him, that is, as soon as it becomes clear that this will be a sad story, this tone disappears entirely, giving way to a serious, objective diction.

The sad story *An Honest Thief* shows no humor in its final version. The original journal version,[22] where the first part of the story is kept in a much lighter vein than the tragic second part, mitigates the shock caused by this study in human abjectness by the warmth and casual, almost amused "human interest" with which the narrator introduces Astafij Ivanovič, who, in turn, tells the story of Emelja, the honest thief. There are some passages which are outright funny. For instance, Astafij's description of the triumphant entry of the Allies into Paris finds Napoleon in person "marching at the very tail end of the parade and shouting, like the rest, 'hurray to the white Czar!'."

A Christmas Party and a Wedding is a rather grim satire. It shows Dostoevskij in a role unusual for him, namely, that of the social critic, armed with the whip of biting sarcasm. One expects Doestoevskij to be ironic, but not in this blunt, ferocious way:

Next to the person who thus participated in the familial bliss of the host, who, incidentally, had five well-fed boys of his own, there was present yet another gentleman for whom I immediately developed a liking. But that one certainly was of quite different calibre. He was a *personage*. His name was Julijan Mastakovič. One could see at first glance that he was the guest of honor, and that his relationship to the host was much the same as that between the latter and the gentleman stroking his sideburns [of whom

[22] It appeared in *Otečestvennye zapiski*, LVII (1848), 286-306, and had two parts: I. *Otstavnoj* and II. *Čestnyj vor*. It bore the title *Rasskazy byvalogo čeloveka* (*Iz zapisok neizvestnogo*). In the edition of 1860, Dostoevskij left out everything but the introductory part (ca. two pages) of part I, as well as the concluding passage of part II (ca. one page), and called the story *Čestnyj vor*.

we have just heard that "the host was patronizing him, but without much gusto"]. Host and hostess were telling him all kinds of compliments, they were courting him, plying him with drinks, coddling him like a babe; they would walk up to their guests to introduce them to him while he, in turn, was being introduced to no one. Toward the end of the party, when Julijan Mastakovič stated that he had rarely spent an evening as agreeably as the present, I noticed tears in the host's eye ... Later, I could not fail to admire the wisdom of my hosts in distributing the presents for the children. The little girl with a prospective dowry of 300,000 received a beautiful doll. There followed the presents for the other children, in descending order, according to the corresponding descent in the rank of their parents. Finally, the last child ... got only a book of tales about the majesty of nature, tears of gratitude, *etcetera*, without pictures or even a single vignette. He was the son of the governess of the host's children. (I, 581)

Dostoevskij, as a rule, is kind to children; here he is in fact slightly hostile: much nicer things could have been said about the little boys than that they were "well-fed". For their parents he has nothing but unqualified contempt. We are soon going to learn that Julijan Mastakovič is everything but a nice person. The point of the irony is therefore that his host, and everybody else at the party, is perfectly sincere in his love and admiration for the great man: the host's tears are genuine! These people are all so perverted that they engage in the vilest sycophancy with natural ease and *con amore*. The ironic description of the book is apparently a side-thrust at the hypocrisy of some works dating from the period of enlightenment (but then still current in Russia) in which the beauty of nature and the wonderful order in this best of all possible worlds were extolled in complete disregard of human misery seen everywhere.

Later, in the scene with the two children, Julijan Mastakovič cuts a most ludicrous figure and, to top it, is insolently laughed at by the narrator, a man of no rank or consequence. Even in the wedding scene he is drawn as a caricature, ridiculous rather than terrifying.[23] A number of rather unpretentious jokes, of a kind not very common in Dostoevskij's work, also appear in this story. For instance, the gentleman with the sideburns is said to be stroking them so assiduously that a bystander might have thought the sideburns were created first and the gentleman attached to them later, just to stroke them. This typically Gogolian "realized metaphor" is not organically connected with the texture of the story, for the gentleman with the sideburns has no part in it. He, his sideburns, and the joke about the sideburns have no function but the immediate comical effect. Altogether, *A Christmas Party and a Wedding* is a satirical sketch

[23] In *A Faint Heart*, he is treated quite differently: with subtle, hidden irony, perhaps, but certainly not as a "funny little round man". Julijan Mastakovič makes still another appearance in Dostoevskij's anonymous feuilleton in the April 27, 1847, issue of *Sanktpeterburgskie vedomosti.*

(rather than a short story) in the less sophisticated journalistic vein. Its relatively unpretentious humor is in accord with such a characterization.

In *White Nights*, the Dreamer makes a heroic effort to attain that most desirable condition of the mind which the German romanticists called *Humor*. He ultimately fails when unable to recover after his grasp for a happiness more intense than that which *Humor* can give ends in dismal failure. In the end, the Dreamer has not only lost his ladylove but his *Humor* has also turned sour on him. But before this happens, he gives us some fine exhibits of a wonderfully serene and kindhearted *Humor* after the fashion of German romanticism at its most mellow. Nasten'ka gaily seconds him.

The gentle self-irony which prevails in the tone of the narrative is sincere and comes from the heart. Our Dreamer is ambitious, proud, and brave only in his dreams. In real life, he makes no pretense to any of these qualities, but humbly makes fun of his own lack of them. When he sees the pretty brunette bent over the grating of a canal, crying, he would like very much to accost her, but dares not, for his "heart is quavering like that of a little bird in the hand of its captor". He does not think much of his courage and good-naturedly admits it. Luck comes to his rescue. An inebriated stranger threatens to molest the young lady, giving the Dreamer (who fortunately happens to be carrying a big stick) a chance to intervene. The inebriated gentleman "takes cognizance of the irresistible argument" *ad baculum* and leaves the scene, accompanied by the Dreamer's profuse expressions of gratitude — pronounced inwardly, for the gallant rescuer is already holding the lady's hand. Note the mock pathos with which the Dreamer welcomes this unexpected stroke of good luck. He is not seeking to conceal the fact that his chivalry has not been a disinterested one.

From here on the Dreamer is firmly in Nasten'ka's little hand. Having foolishly promised the girl not to fall in love with her (naturally he is already in love, and deeply so), he soon finds himself in the unenviable position of her confidant who is *nolens volens* promoting the fortunes of his rival. He realizes the irony of this situation, but accepts it gracefully, with only a tinge of melancholy added to his humor. And so we see him trying to convince Nasten'ka that it is the most natural thing in the world for a young man, very much in love, not to have bothered to get in touch with the girl he loves, although in town for at least several days; also, that it is perfect *bon ton* for that girl to write him a torrid love letter nonetheless (II, 38). The irony is a twofold one. For one thing, the Dreamer's arguments are neither logical nor convincing. But Nasten'ka, who, of course, wants to believe what he is saying, is convinced and grateful for such marvellous advice. And for another, the Dreamer is using all his eloquence and dialectic skill trying to make her do what serves his own interests worst.

The Dreamer's confession is everything that the Serapion Club could have sought in a creation of one of its members: it is inspired, coming straight from the Dreamer's heart; yet it is poetic (take, for instance, that beautiful encomium of fantasy, I, 22-23),[24] ironically ambiguous (in the end we do not know if the whole thing has been an apology or an indictment of daydreaming), and genuinely humorous. Time and again, the Dreamer interrupts his own pathetic recital with ironic comments, such as (from one of his favorite daydreams): "How innocent, how pure was their love, and how (yes, Nasten'ka, you've guessed it) how wicked were the people around them!" (II, 26). The interpolation ought to make it clear even to naïve Nasten'ka that he is laughing at himself. "How innocent, how pure was their love" may be an impish "side sweep" aimed at pure love *à la George Sand*, for we learn that "She" is married, of course to an elderly, taciturn, bilious aristocrat who has no understanding for her tender feelings, yet ready to fly into "His" arms the moment she is free, and perhaps even before. "He" is of course the Dreamer himself.

Toward Nasten'ka, the Dreamer is all tact and tender attention. He lets humor take care of itself, and it does. Nasten'ka's healthy egoism and resolute determination to be happy cannot help producing some funny effects. For instance, when she has accepted the fact that her fiancé is apparently not going to come after all, to transfer her affection from him to the Dreamer (second choice, but available) is for her the work of a moment:

> I love him; but this will pass, it must pass, it can't but pass; it is passing already, I feel it ... who knows, maybe it will pass already to-day, because I hate him ... (II, 52)

The Dreamer is not so ungentlemanly as to take notice of the speed of her recovery, but the reader notices it anyway. This joke, an innocent, good-natured one, is on Nasten'ka, and there are others where it is she who draws a laugh. But much more often Nasten'ka is seen making fun of someone else — the Dreamer, her grandmother, Fekla the maid. For example, Nasten'ka likes to read novels. Her grandmother keeps asking her whether these novels are indeed entirely "moral", for "if they are immoral, Nasten'ka, you shouldn't be reading them". When Nasten'ka wants to know the reason why, she gets this answer:

> Ah, they're all about young gentlemen seducing innocent young girls, and how they carry them off from their parents' homes, pretending they want to marry them, and how they later abandon those unfortunate girls to their fate, so that they must perish in the most lamentable fashion. (II, 33)

The roguish Nasten'ka is paraphrasing the words of her grandmother so as to

[24] The Dreamer starts his eulogy with a reference to Žukovskij's poem *Moja boginja (podražanie Gëte)*, where the "goddess" is, of course, fantasy.

make them seem particularly naïve, although well meant. The point is that grandmother's example of an "immoral" book is not a very felicitous one, for although an immoral act (seduction and elopement) is described there, it is still a most "moral" book, in fact, it is a book with a moral. A moral, moreover, which Nasten'ka might very well apply to herself. Which she knows and cheerfully ignores.

The Dreamer is of course the principal target of Nasten'ka's wit. Since she has a liking for him, her little "digs" at his rhetorical talents and tendency to be carried away are, well, almost compliments. In the meantime, however, the situation created by Nasten'ka's trustful acceptance of the enamored Dreamer as her confidant makes for some "laughter through tears" for the Dreamer, for instance, when Nasten'ka finally admits that she is a little disappointed at his not having fallen in love with her. Little does she know — poor Dreamer!

The irony of the whole situation ultimately hinges on the ambiguity of the word "love". When the Dreamer finally confesses that he loves her, Nasten'ka cuts him short: "Well, and so? Really, what of it? Why, I've known for a long time that you love me, only I thought, well, I thought you simply loved me, just simply ... oh dear, oh dear!" She seems genuinely surprised that the young man who has been virtually at her feet ever since they first met should be in love with her, instead of loving her "simply". The point is that, up to now, she has wanted the Dreamer's love to be brotherly, selfless, pleasant — what she calls "*simply* love". Only when the Dreamer is about to take her fiancé's position is she willing to give his "love" a different meaning. The moment her fiancé is back she immediately substitutes the old meaning: "I want him [her fiancé] to see how we love each other!" she tells the Dreamer when they believe her fiancé to be approaching them and the Dreamer is about to withdraw hastily. She presumes that both young men will be delighted to form a "triangle" around her. The fact that there may be other viewpoints beside hers does not exist for her. Sometimes the poor Dreamer doesn't know whether he ought to laugh or cry at so much naïve egoism.

In Nasten'ka's farewell letter the irony bred by the ambiguity of the word "love" is driven to the limit: Nasten'ka still "loves" the Dreamer, but is going to marry the other young man next week! What had been a game of semantic hide-and-seek now becomes cruelty. Yet there is so much sweet naïveté about Nasten'ka's possessiveness and self-assurance (she is trying to keep *both* men and seems confident of success) that even this cruelty elicits a smile.

Nasten'ka is one of the rare heroines of world literature in whom the charm and beauty of the *primadonna* are combined with the wit and humor of the *soubrette* and the touching simplicity of the *naïve*. A wonderfully gay and lovely figure she is. The genuine humor which imbues it makes Nasten'ka a truly romantic figure — a *Prinzessin Brambilla* in real life, without masks, or theater, or Roman carnival.

Very much unlike most of Dostoevskij's works of the early period, *Netočka Nezvanova* is written in a serious, objective tone almost throughout. The use of verbal humor and of irony is limited to a minimum. There are, however, a few humorous scenes in Part II, mostly involving Sir John Falstaff, the bulldog. Here, humor of the situation is combined with humor of the word. The latter consists mainly in referring to the dog in terms of human character traits and social relations. Falstaff is said to be "proud, arrogant, and status-conscious" (*spesiv, gord i ambiciozen*, II, 158). He loves no one, but insists on being respected by every member of the household. He gets this respect, mixed with a good deal of fear — until the "important" spinster aunt arrives and Falstaff is "formally excluded from the upstairs suite of rooms" (*ibid.*). The climax of the Falstaff intermezzo, where we see naughty little Katja open the door for Falstaff so he will run upstairs and scare the old lady, is kept in a tone of mock seriousness and feigned excitement:

> Falstaff flew by us like a cannonball. Mme. Léotard shrieked in terror, but Falstaff had already reached the door of his secret dreams; he rammed it with both paws and, having failed to push it open, gave out a fearful howl. The spinster responded with desperate wails of her own. But already whole legions of enemies were streaming to the scene ... and Falstaff, Falstaff the fierce, was ingloriously dragged from the battlefield on a lasso, with a muzzle thrown dexterously over his terrible jaws and all four feet tied tightly together. (II, 160-161)

Earlier in the scene it was stated that Falstaff "had crossed his Rubicon". The whole episode is described in terms of a battle scene. The humor is certainly of a most innocuous nature. It is not very "Dostoevskian", to be sure, but then again it does fit the image of the "lady writer", whom Dostoevskij is impersonating in this novel, very well. Other humorous passages, found occasionally throughout the novel, are of the same nature: quite tame, not too witty, but giving evidence of a lively mother wit, well becoming so wise and talented a woman as the narratress of *Netočka Nezvanova.* Dostoevskij's usual "trademark", his irony, is absent. With good reason, for it would hardly fit the image of the narratress. Occasional flashes of ironic diction must be evaluated much like the occasional colloquialisms found in the narrative: after all, the narratress is only human, and has her likes and dislikes. She certainly dislikes Princess Kh–aja, and the scene (II, 127) in which the latter has invited all her friends to "show off" her little orphan, as well as a few other references to that personage, are in a sardonic vein. When she reports that the proud Princess "would twice every week, on appointed days, dutifully betake herself upstairs to pay her respects to her aunt" (II, 129) her *Schadenfreude* is quite obvious. There is also a good deal of irony in two or three references to the pious spinster aunt, and in one or two passages about Petr Aleksandrovič, Netočka's enemy in Part III. A few times we run across an instance of good-natured, gentle humor, of the type which dominates the narrative of *White Nights*. Some examples of it can be found in the description of the few blissful days (and nights!) when Katja and Netočka have finally found each other and are happily in love. The narratress treats the infatuation of two schoolgirls, which will last only a few days anyway, with a half amused and half wistful smile.

Finally, there is Dostoevskij the feuilletonist, trying to come up with some humor "after the French fashion". *A Novel in Nine Letters* is an amusing trifle, which Dostoevskij claimed to have written in one night.[25] The correspondence is carried on by two professional "cardsharpers" and reveals a string of extremely trite farcical situations. The verbal humor is equally unpretentious, resting entirely upon the contrast between the solid, substantial air the two crooks are assuming throughout and their nefarious schemes, which can be easily reconstructed by reading between the lines.

The somewhat more substantial *Polzunkov* features as its hero and narrator a type which Dostoevskij described often in his later works, namely, the "voluntary buffoon". However, the psychology of the voluntary buffoon, analyzed in some detail in the introductory part of the story, is only superficially integrated into Polzunkov's narrative — less so than, for instance, in the case of Zimovejkin in *Gospodin Proxarčin*. The reader's attention remains focused on the plot, which is anecdotal and contrived. What verbal humor there is in the story is based on the same effect as that in *A Novel in Nine Letters*. Fedosej Nikolaevič and his wife, obviously the most shamelessly depraved, viciously predatory people, give themselves the air of good-natured *bonhommie* ("Old Russian" style), and of unsophisticated, homely integrity. The farcical transparency of the deceit accounts for much of the comical effect. The gullible Polzunkov, for whom nobody must feel sorry since he is as much of a crook as his "boss", contributes to this effect by the hyperbolic stupidity which he displays throughout. The whole thing, while amusing, is quite unworthy of Dostoevskij's genius. It reminds one of Gogol at his most uninspired, say, in *The Carriage*.

The two stories *Another Man's Wife* and *The Jealous Husband*, combined into one story, *Another Man's Wife and the Husband under the Bed*, in the edition of 1860, are even less pretentious. The humor of these two episodes from the life of a jealous husband is almost entirely one of the situation. Thus, in the second episode, Ivan Andreevič, once more in hot pursuit of the voluble Glafira

[25] His letter to his brother Mixail, dated November 16, 1845 (*Pis'ma*, I, 84-85).

Petrovna, enters the wrong apartment and, like a bombshell, bursts into the bedroom of a young lady whom he has never seen before. Before he can withdraw, her husband, an elderly "Excellency", appears on the scene, and the jealous husband, much to his own surprise, finds himself under the lady's bed. To his even greater surprise, he meets another, younger man there, *etc.* The dialogue is quite entertaining, but in a very superficial fashion. Its highlight: the transformation of the blustering, dignified Ivan Andreevič into a harmless, smiling, fawning buffoon, when he suddenly faces His Excellency. All in all, it is not such a bad job, considering that this genre is really a French monopoly. What is *piquant* in French, is only too easily vulgar in almost any other language, including the Russian. Obviously Dostoevskij is quite unoriginal in this genre, and is merely trying to transplant into Russian soil some offshoots of the frivolous light wit in which only the French are really at home. Yet I do not think that any Russian ever did this any better than the young Dostoevskij — in prose, that is. Puškin, of course, did it considerably better in verse, in *Graf Nulin*, for instance.

To summarize: the young Dostoevskij is as versatile as a humorist as he is in every other respect. His humor in *Poor Folk* spans a wide range of different comical effects: literary parody (and travesty!), satire, comic situations, ordinary *Schadenfreude*, subtle insinuations, slapstick comedy, unintentional (and intentional!) buffoonery on the part of the hero, deflation of "important" personages, plus countless other effects, yet all of them integrated by the basic and ever-present theme of Devuškin's simple soul struggling with the complexities of the world in which he lives. *Poor Folk* is a *roman comique* in the epic tradition of *Don Quixote*, *Joseph Andrews*, and *The Pickwick Papers*.

In *The Double* we have most of the same shades of humor — plus a grotesque puppet theatre, a veritable counterpoint of mocking echoes, and the persistent irony of the narrator. The last mentioned is, I think, the integrating factor of the humor in this novel. What is lacking in *The Double* (as compared with *Poor Folk*) is the hero's own sense of humor: Goljadkin is not an inherently funny figure. *The Double* is, I believe, best characterized as a tragicomic *mystère*.[26] It resembles most of Dostoevskij's mature works in that it contains strong elements of drama. The humorous effects in *The Double* are best appreciated if the novel is conceived in a "dramatized" form and "visualized".

Gospodin Proxarčin is the irreverent, cynical, nihilistic treatment of a serious

[26] The conception of the Dostoevskian novel as a *mystère* (*misterija*) is developed by Leonid Grossman, *Put' Dostoevskogo* (= *Sobranie sočinenij v pjati tomax*, Vol. II) (Moscow: "Sovremennye problemy", 1928), p. 8. It is accepted, with some reservations, for *The Double* by Baxtin, *op. cit.*, p. 156.

theme, a story in which everybody and everything is made fun of — including the tragic insight which is the theme of the story. *Gospodin Proxarčin* is a tragic *grotesque*.

There is some delicate, warm humor in *A Faint Heart*. It clears the way for the pathos of compassion. *White Nights* is another fireworks of a variety of humorous effects: light banter, irony, parody, comic situations, ironic ambiguity. I see *White Nights* as a wonderfully charming, gaily melancholy *comedy of character*.

A Christmas Party and a Wedding and *Polzunkov* contain some savage social satire, which is, however, enough *outré* to be funny. Finally, there is the young Dostoevskij's "humor after the French fashion". What more, how much more variety could one expect from a young writer within a span of three years?

BYRON'S *DON JUAN* AND PUŠKIN'S *EVGENIJ ONEGIN*: THE QUESTION OF PARALLELISM

WALTER N. VICKERY

University of Colorado

No consensus has as yet been reached by scholars as to the duration, intensity or the precise nature of Byron's influence on Puškin.[1] While many factors have combined to obscure this problem, three are particularly relevant to the present study: 1) the insistence by some scholars that at a particular period Puškin cast off romanticism in favor of realism has frequently been tied in with his alleged rejection of Byron; 2) Byron's influence on Puškin coincided with and in some cases overlapped with the influence of other Western writers;[2] 3) insufficient attention has, in my view, been given to the common literary tradition shared by both men. In a recent study I attempted to stress this common literary ancestry in order to show: 1) that Byron's influence was no mere chance phenomenon, no rapidly outgrown phase, and; 2) that — in spite of important ethnic, social and other differences — there was in the development of the two men a certain parallelism.[3] Clearly, if parallelism — as a concept applied to Byron and Puškin — is to be made palatable, viable and convincing, then a great deal depends on establishing and defining the relationship between *Don Juan* and *Evgenij Onegin*. That a relationship exists is not normally denied. As to its nature, considerable work has already been done on this from varying viewpoints and, often, with differing conclusions. The aim of the present study is to contribute to the overall discussion by focusing attention on: 1) certain examples of parallelism in plot, situations, characters and social commentary, and 2) a certain similarity, related to the problem of genre, in the *method* by which these

[1] Reference will be made in appropriate places to Leon Stilman, "Problemy literaturnyx žanrov i tradicij v 'Evgenii Onegine' Puškina", *American Contributions to the Fourth International Congress of Slavicists* ('s-Gravenhage: Mouton & Co., 1958), pp. 321-367; but I wish at this point to acknowledge my overall indebtedness to this perceptive article. Without making its author responsible for any of the views expressed in the present study, I should like to put on record that his article has served as a powerful stimulus and help in my work on *Evgenij Onegin*.

[2] E.g. Chénier, Chateaubriand, B. Constant.

[3] W. N. Vickery, "Parallelizm v literaturnom razvitii Bajrona i Puškina", *American Contributions to the Fifth International Congress of Slavists* (The Hague: Mouton & Co., 1963), pp. 371-401.

two poems produce their *esthetic impact* and, even, a certain similarity in the *nature* of the *esthetic impact* produced.

Both *Don Juan* and *Evgenij Onegin* are new in genre and hybrid in genre. Both lean in form to the tradition of the epic — more specifically the comic epic or the humorous verse tale, while at the same time both are related to the novel.[4]

In the field of comic epic — and the closely related humorous verse tale — it is not difficult to establish a literary ancestry common to both Byron and Puškin. "Beppo" and *Don Juan* are nineteenth-century descendants of the Italian Renaissance comic epic of Ariosto and Pulci, of Boiardo and Berni. The immediate bridge with the Italian Renaissance was provided by John Hookham Frere's "The Monks and the Giants", written in *ottava rima.* At the same time Byron reckoned among his literary forbears in this field such writers as Voltaire, Lafontaine and — in his native tradition — Dryden, Pope and Chaucer. Puškin shared part of this tradition. His "Ruslan i Ljudmila", written before he came under Byron's spell, reflects the influence of Ariosto and of Voltaire's *La Pucelle*, and — in his native tradition — of Majkov and Bogdanovič, the latter of whom was in his turn much influenced by Lafontaine. Thus, in this field a crisscrossing of the same or similar literary influences can be observed, and its significance in terms of the present thesis of parallelism should not be underestimated.

It is, however, on the relationships of *Don Juan* and *Evgenij Onegin* to the novel that I wish here to focus attention. For it is when we consider the types of novel that helped mould the conception of *Don Juan* and *Evgenij Onegin* that the paths of the two writers at first sight appear to diverge.

In Byron's case the major influences are Voltaire, Fielding and, in some respects, Sterne. *Don Juan* belongs to the picaresque tradition of *Candide* and *Tom Jones*; as one critic has remarked, "the real model of *Don Juan* is the picaresque novel, the great catchall of narrative and reflection, subject to no law but the author's desires".[5] In accordance with the picaresque tradition, Byron moves his hero from episode to episode and from one place to another.

Puškin, it scarcely needs to be said, knew Voltaire, Fielding and Sterne.[6] But the influence of these three writers was — as far as the plot of *Evgenij Onegin* is

[4] It may be noted that around the period when Byron and Puškin wrote there was a tendency for these two genres — comic epic and the novel — to coalesce. Puškin gave *Evgenij Onegin* the subtitle of a "novel in verse"; in her *Byron's Don Juan* (New Brunswick, 1945), p. 44, E. Boyd — without any thought of Puškin — speaks of *Don Juan* as a "novel in verse"; and considerably earlier, in his forward to *Joseph Andrews* Fielding described the picaresque novel as "comic epic in prose".

[5] E. Boyd, *op. cit.*, p. 34.

[6] Sterne's influence on *Evgenij Onegin* has been treated by V. Šklovskij, "'Evgenij Onegin' (Puškin i Stern)", *Očerki po poètike Puškina* (Berlin: Epoxa, 1923), pp. 194-220.

concerned — far outweighed by that of a different type of novelist, by that of such authors of sentimental-romantic novels as Rousseau (*La Nouvelle Héloïse*)[7] and Benjamin Constant (*Adolphe*).[8] And, as opposed to Byron, Puškin confines his plot to one central episode and to one country — his native Russia.

It is the general contour of the plot of *Evgenij Onegin* and, in particular, Puškin's confining his plot to one episode and his own country that have led commentators to insist that *Evgenij Onegin* truly is a "novel in verse", the first genuine Russian novel in the realist tradition of Turgenev and Tolstoj — and therefore basically different in genre to *Don Juan*. All this notwithstanding, I would like to suggest that the differences between the two works are not as fundamental as might at first sight appear: 1) Byron's *Don Juan* shows signs of shifting from the picaresque tradition to a background which to some extent parallels that of *Evgenij Onegin*; 2) the *method* by which *Evgenij Onegin* produces its *esthetic impact* more closely resembles Byron's method than that of Rousseau or Benjamin Constant, let alone Turgenev or Tolstoj.

1. *Parallelism in Plot, Situations, Characters and Social Commentary.* Starting out in the picaresque tradition, Byron moves his hero through a rapid succession of episodes (the Julia affair in Spain, the shipwreck, Haidée's island, the harem, the battle of Ismail, and the confrontation with Catherine the Great in Russia). But after the tenth canto which brings Juan to England, Byron's native land, there is no further migration, and at the end of the sixteenth canto — with the poem unfinished — Juan is still in England. A comparison of *Evgenij Onegin* with these last six cantos of *Don Juan* (when the action is confined to one episode and one country) reveals definite parallels: 1) *Evgenij Onegin* starts with a description of Onegin's social life in Petersburg based largely on the exiled Puškin's recollections; in cantos XI-XII Byron describes Juan's social round in London in similar vein; 2) in both poems the scene then shifts to the country, and (notwithstanding substantial differences in background and plot) there occurs in both poems a house party, at which the guests, bearing allegorical names, in the tradition so often honored in the eighteenth century, are described satirically; 3) Aurora Raby's superiority to her environment (she is calm, serene, retiring, thoughtful, somewhat sad, etc.) is reminiscent of the youthful Tatiana, while Aurora's poise which discomforts Juan is reminiscent of Tatiana's poise when, a married woman now in the last chapter, she on several social occasions leaves Onegin tongue-tied or abashed; 4) with the entry into the story of Aurora Raby (let alone Lady Adeline, whose complicated personality belongs not to

[7] Cf. L. Stilman, *op. cit.*

[8] Cf. A. A. Axmatova, "'Adol'f" Benžamena Konstana v tvorčestve Puškina", *Puškin — Vremennik puškinskoj komissii* (Moscow-Leningrad, 1936), I, 91-114.

the picaresque tradition but to the tradition of the social novel) we have for the first time in *Don Juan* a heroine on a par with the hero; 5) the conception of Aurora Raby, as of Tatiana, clearly owes something to the tradition of Rousseau, whose influence has up to this point not been in evidence in *Don Juan* (with, of course, the important exception of the incident involving Haidée, who — it may be noted — is compared to Aurora); 6) Juan, more sophisticated now than at the outset of the poem, somewhat jaded and not overly susceptible to female charms, is nevertheless moved by his meeting with Aurora in much the same way as was Onegin by his meeting with Tatiana (Onegin's reactions are stronger if we include the omitted draft variants).[9] Thus, in several ways, and without wishing to push the comparison too close, from the eleventh canto on *Don Juan* turns onto a course roughly parallel with that of *Evgenij Onegin*; the picaresque tradition no longer dominates, a stable realistic background is etched in, complete with secondary characters, Rousseauesque overtones make their appearance in the characterization of the main female protagonist, and there are intimations that — had the poem proceeded further — a love story set on a conventional social background was in the making.

How Byron would have completed *Don Juan*, whether he would have left his hero permanently in England, is open to doubt. (He is on record as having planned to bring him to his death in the French Revolution[10] — which would, incidentally, have established another parallel of a sort with Puškin, whose fragmentary tenth chapter and the evidence surrounding it would seem to indicate that Onegin was, had Puškin continued the poem, to have become involved

[9] Puškin, *Polnoe sobranie sočinenij* (Izd. Ak. Nauk SSSR, 1937), VI, 307-308:

В постеле лежа — наш Евгений
Глазами Байрона читал
Но дань вечерних размышлений
В уме Татьяне посвящал —
Проснулся ⟨он⟩ денницы ране
И мысль была все о Татьяне
Вот новое подумал он —
Не уж-то я в нее влюблен
Ей богу это было б славно
[Себя] [уж] то-то б одолжил
Посмотрим — и тотчас решил
Сосед⟨ок⟩ навещать исправно
Как можно чаще — всякой день
[Ведь] им досуг а нам не лень

[10] Letter to Murray, February 16, 1821, *The Works of Lord Byron*: *Letters and Journals*, ed. R. E. Prothero (London: Murray, 1901), V, 241-245. In the same letter Byron speaks of making Juan "a cause for a divorce in England". For speculation on the possible continuation of *Don Juan*, also for an excellent discussion of the changes in approach and characterization in the "English" cantos cf. Edward E. Bostetter, *The Romantic Ventriloquists* (Seattle: University of Washington Press, 1963), pp. 241-253.

with the Decembrists).[11] Meanwhile, it can be stated that the eight completed chapters of *Evgenij Onegin* have a good deal more affinity with the last six cantos of *Don Juan* than with the first ten.

It should be pointed out, furthermore, that this affinity cannot be simply ascribed to "borrowing" on Puškin's part. The question of "borrowing" is a complicated one, and I do not wish to go into it in detail here. But since the very idea of "borrowing" is in conflict with the idea of parallelism, a few comments are in order. A certain amount of borrowing can in general be assumed. Puškin is in some degree indebted to Byron for the whole idea of a "novel in verse". The technique of the author's digression he in large measure owes to Byron — at least, the use of this technique in the context of a novel in verse. There are many lines in *Evgenij Onegin*, particularly those involving such subjects as women, love, friends, the social round, and to some extent literature, which — if space permitted — could be set side by side with lines from *Don Juan* and seen to bear a strong resemblance[12] (though in some instances the thought is new with neither Byron or Puškin, but goes back to such predecessors as Voltaire). More specifically, the idea of the house party and the house guests with their ridiculous allegorical names may also have been taken from Byron, though it certainly was not new with him.[13] Something of Aurora Raby's social poise may have gone into the Tatiana of Puškin's eighth chapter. And surely other examples could be cited. But the point I wish to stress here is that the affinity between *Evgenij Onegin* and the last six chapters of *Don Juan* cannot in the main be accounted for by the idea of "borrowing". For the fact is that by the latter half of November 1825 Puškin had still only read the first five cantos of *Don Juan*,[14] and by September 14 he had basically finished the fourth chapter of *Evgenij Onegin*.[15] Thus, Onegin's life in Petersburg had been described; he had gone to the country; Tatiana, Ol'ga and Lenskij had been introduced; Tatiana had written her love letter; Onegin had lectured her in the garden; and the idea of a name-day party for Tatiana had been mentioned — all this while Juan was still in the harem. Therefore, the similarities noted above between cantos XI-XVI of *Don Juan* and *Evgenij Onegin* are based — at least in part — on the common literary heritage shared by the two authors and on their literary evolution along roughly parallel lines. And from the standpoint of genre, the distance separating Puškin and the author of the last six cantos of *Don Juan* is not as great as it appeared initially, when we spoke of Puškin's leaning toward

[11] Cf. B. V. Tomaševskij, *Puškin* (Izd. Ak. Nauk SSSR, 1961), II, 200-244.

[12] For some examples cf. L. Stilman, *op. cit.*, 338-342; also W. N. Vickery, *op. cit.*, 392-396; but many more examples could be given.

[13] Cf. e.g. T. L. Peacock's *Melincourt*.

[14] Cf. his letter to Vjazemskij of November 1825, Puškin, *Polnoe sobranie*, XIII, 243-244.

[15] Cf. his letter to Katenin of September 1-14, 1825, Puškin, *Polnoe sobranie*, XIII, 224-225.

the Rousseauesque sentimental-romantic novel and of Byron's leaning toward the tradition of the picaresque novel.

2. In another and far more important way the two poems resemble each other: in the *manner in which they produce their impact on the reader*. Every poem and every work of literature may be said to have a *fundamental theme* (what Croce called the *motivo fondamentale*), which is its own, which differs from that of every other poem or work of literature, which *is* the poem or the work of literature in question. In this sense *Hamlet* is *Hamlet* and *Anna Karenina* is *Anna Karenina*. In this sense every work of literature is unique and its impact is unique. At the same time there do exist similarities and differences in the various ways in which works of literature produce their impact; and these similarities and differences frequently relate to the time in history at which the works were written and also to the question of genre. I shall try to show that in this latter sense — *in the manner in which they produce their impact* — *Don Juan* and *Evgenij Onegin* have a good deal in common.

First *Don Juan*. The *fundamental theme* of *Don Juan* — as of every other work — is made up of the sum of the component parts. It is fair to say that every incident, amusing or tragic, every character, good, bad or indifferent, every comment from the author — these and other elements all make their contributions to the overall impression, the overall impact. But how important relatively are these different contributions? What is the work's esthetic organizing principle which will determine the hierarchy of these many different elements and which will determine, so to speak, the point of esthetic impact on the reader or the reader's esthetic focus on the work? What position in the hierarchy is occupied by the individual incidents and the individual characters in *Don Juan*? Is the reader deeply concerned, for example, about whether Gulbeyaz will seduce Juan? Or whether Juan will seduce Julia? Is he deeply concerned about whether Juan will escape from the harem? Is he concerned about Juan's health in Russia? Is he even very deeply concerned — or for long — about Haidée's death? My impression is that though the reader may feel a certain suspense as to the outcome of the various episodes, even a certain concern for the individual characters, his feelings are not as deeply involved as they would be in, for example, the fate of Tolstoj's Anna Karenina. If I am right, then this lack of deep involvement is partly due to the humorous tone in which most of the episodes (excluding the Haidée episode and, perhaps, the shipwreck episode) are narrated. But it is also very much due to the omnipresence and domination of the author's personality. The author constantly intrudes, comments, discussed his own and others' experiences, social mores, political issues, literary problems, even the way he intends to manipulate his characters. And in so doing he considerably weakens in the reader the illusion that real events and real people are involved; he offers, as it were, a dream from which the reader can at any moment extricate himself merely by turning over on the other side. This does not weaken the esthetic impact of *Don Juan*; it simply changes it. It means, in effect, that events and characters, though they do contribute very substantially to the impact of the poem, are not the primary instruments in setting its tone; the tone is set and the focus is created by the literary personality of the author. That this is so may be seen from the fact that although, as noted, the incidents are mostly narrated in humorous fashion, the prevailing mood of the poem is by no means one of gaiety. "*Don Juan* begins in fun", one critic has remarked, "but it ends in bitterness and sadness".[16] It ends, to be sure, with the humorous incident involving "the phantom of her frolic Grace — Fitz-Fulke". Yet the prevailing mood is correctly characterized as one of "bitterness and sadness". Gone is the devil-may-care wit and sparkle of the first canto in Spain. We still hear the laughter, hear it to the end, but we hear it muted by somber, tragic tones. And the responsibility for this lies with the literary role played by the author, who acts as the esthetic organizing principle, the prism through which events and characters are seen.

Edmund Wilson, whose assessment of *Don Juan* seems unjustifiably harsh, has written that "essentially it is a drunken monologue by a desperately restless, uncomfortable man, who does not know what is the matter with him or what he ought to do with himself, who wants to tell stories about other things or to talk about himself in such a way as to be able to laugh and curse and grieve without looking too closely into anything. Byron's achievement, certainly quite remarkable, is to have raised the drunken monologue to a literary form."[17] Without agreeing with Wilson's low assessment and leaving aside the question of drunkenness, it should be recognized that his remarks contain one very important truth; for in the final analysis it is Byron — notwithstanding all the episodes and characters and objective descriptions of society — who is the *lyric hero* of this poem, the tragic theme of which is the passing of youth and the onset of — whatever follows youth.

So much for *Don Juan*. All this may seem a long way removed from *Evgenij Onegin*. If we are to look for the organizing principle in *Evgenij Onegin*, it must surely be found in the events that unfold and in the characters that participate in these events? It must be found in Tatiana's unrequited love for Onegin and in Onegin's unrequited love for Tatiana? It must be found in the friendship of Onegin and Lenskij and in the foolish duel that led to Lenskij's death?

16 Boyd, *op. cit.*, p. 31.

17 Edmund Wilson, "Evgeni Onegin", *The Triple Thinkers: Ten Essays on Literature* (New York: Harcourt, Brace & Co., 1938), p. 45.

It must be found in the *mal du siècle* of Onegin and the aimlessness of his existence? Surely, since *Evgenij Onegin* is a "novel in verse", these events and characters must be taken seriously, at face value, as real events and real people? The answer is, I believe, that of course *Evgenij Onegin* is a "novel in verse" — if for no other reason than because Puškin gave it that subtitle; but that there are many different types of novel and to label a work a novel does not necessarily tell us very much about the manner in which it produces its esthetic impact. A comparison of *Evgenij Onegin* with its apparently nearest kin in the field of the novel proper reveals a basic difference in the *narrative method.* For the narrative method of the epistolary novel, as exemplified by *La Nouvelle Héloïse*, insures — automatically and by definition — that the author remain entirely behind the scenes and that the characters writing the letters which advance the narrative actually existed — or so we are asked to believe. And to an equal degree the first-person narrative of *Adolphe* (a sort of confession allegedly found and published by someone else) demands that the reader believe in the reality of the hero and of the characters he comes in contact with and describes. But the narrative method of *Evgenij Onegin* is entirely different. It is very similar to that of *Don Juan.* For once again the author plays a decisive role — not, of course, as the one-time friend of Onegin, but as the narrator who constantly intrudes, comments and — more importantly — discusses the way he manipulates the characters. If in the characterization of Tatiana and Onegin *Evgenij Onegin* is related to *La Nouvelle Héloïse* and *Adolphe*, it is on the issue of narrative method far more closely related to *Tristram Shandy* and *Tom Jones* — the connecting link being, of course, *Don Juan.* The following comment on the narrative method in *Evgenij Onegin* seems very appropriate: "Especially significant is the highly complex author-image and the narrator's interventions which reveal the creative process and remind the reader that he is dealing with a work of imagination. What lacks in *Evgenij Onegin* is the basic assumption of realistic literature, namely that the narrative will be accepted as an equivalent of actual experience."[18] This does not mean — any more than it did in the case of *Don Juan* — that the esthetic impact of *Evgenij Onegin* is weakened; it is merely changed. As in *Don Juan*, there is nothing which does not contribute to the overall impact of the poem. The metaphor of the revolving seasons has its important role to play.[19] Onegin, Tatiana, Lenskij, Ol'ga and others play, assuredly, their vital part. But it is not in any of these that the esthetic focus, the center of gravity, the point of esthetic impact should be sought; it should be sought rather in the *lyric hero* who, no less than

[18] Stilman, *op. cit.*, p. 366.

[19] For the function of the revolving seasons cf. R. F. Gustafson, "The Metaphor of the Seasons in *Evgenij Onegin*", *Slavic and East European Journal*, VI, 1 (Spring 1962), pp. 6-20.

his counterpart in *Don Juan*, constitutes the dominating force, the esthetic *organizing principle* of this novel in verse.[20] The kernel of the poem lies not in Onegin's rejection of Tatiana nor in Tatiana's subsequent rejection of Onegin; nor yet in Lenskij's death — either in itself or for its effect on Onegin. These and other things are of importance in the general conception of the poem, but they are not central to that conception. Rather they contribute to a feeling — which is the very personal feeling of the lyric author-hero — that time passes by and is irretrievably lost, there is no return, no recapturing what is gone.[21]

Recognition of the fact that the author of *Evgenij Onegin* is also its lyric hero and organizing principle facilitates an understanding of certain problems which arise in connection with some of the textual variants and omitted stanzas. It is well-known to students of Puškin that an examination of certain of the textual variants, omitted in the final version of the poem, leaves the impression that Puškin had moments of vacillation as he went about the business of creating the individual characters. When introduced in the second chapter, Lenskij appears almost hopelessly idealistic and naive, the attitude towards him would seem to be one of condescension; yet several stanzas, omitted at this point, reveal the author's sympathy for and, in large measure, approval of Lenskij's youthful idealism.[22] Originally it was Ol'ga, not Tatiana, whose relationship with her nurse was described as close.[23] And an examination of the draft variants shows Puškin several times hesitating as to how much distance he wishes to put between himself and Onegin and what sort of person Onegin will be.[24] Since *Evgenij Onegin* was written over a period of years, it seems that it was inevitable

[20] Stilman, *op. cit.*, p. 330, makes essentially this point when he speaks of the "lyric author". Something of the same approach may be found in V. Šklovskij, *op. cit.*, p. 211, where he remarks: "The real subject of *Evgenij Onegin* is not the story of Onegin and Tatiana, but a play (*igra*) on this plot." And when Šklovskij asks (*op. cit.*, p. 214): "To put it crudely, did 'Puškin' shed tears over Tatiana, or was he joking?" he is again clearly indicating his feeling that the esthetic focus of *Evgenij Onegin* is beamed not on the main characters, but elsewhere. Where, however, Šklovskij's view differs essentially from mine (and, I believe, Stilman's) is in his contention that *Evgenij Onegin* is really a parody on the novel, a joke perpetrated by the author.

[21] This point may be made clearer by comparison with Čajkovskij's opera by the same name. In the opera (music apart) our involvement is directly concerned with the characters involved; the center of esthetic gravity has shifted from the lyric author-hero to Onegin, Tatiana and Lenskij.

[22] The omitted stanzas begin:

1) Не пел порочной он забавы ...
2) Певцы слепого упоенья ...
3) Пустыми звуками, словами ...
4) Но добрый юноша готовый ...

Puškin, *Polnoe sobranie*, VI, 270-272 and 282-283.

[23] Cf. Puškin, *Polnoe sobranie*, VI, 565-567.

[24] Cf. Ju. M. Lotman, "K èvoljucii postroenija xarakterov v romane 'Evgenij Onegin'", *Puškin: Issledovanija i materialy* (Izd. Ak. Nauk SSSR, 1960), III, 131-173.

that attempts would be made to explain these "vacillations" in terms of Puškin's changing views.[25] I would not wish to imply that such attempts have been unfruitful. But the problems raised by these vacillations become, I believe, more easily acceptable, less weighty, and fall more smoothly into place if we bear in mind that Puškin himself is the lyric hero of the poem and that Onegin (in spite of Puškin's disavowal), Lenskij and Tatiana (who at the end is identified with his Muse) are all very closely related to the author's lyrical emotions. They are all, if you will, characters in their own right; but they are not quite free of their author, the puppeteer's manipulating strings still show.

Recognition of the author as the lyric hero of *Evgenij Onegin* has a bearing on one further problem connected with the construction of the poem. It has been pointed out before that a change in both mood and content occurs in *Evgenij Onegin* early in the poem. After the satirical, light-hearted description of society life in the first chapter, we are plunged into two love affairs — Lenskij and Ol'ga, Onegin and Tatiana — and a change in mood is perceptible. This has been described as a transition from "a satirical poem to a social novel of everyday life", involving a switch in the author's artistic intent.[26] It has, on the other hand, been pointed out correctly that the first two chapters were written in rapid succession — and this fact has been used to reject the notion of any switch in artistic intent, the underlying thought behind this rejection being that a switch in intent would somehow be artistically a blemish.[27] But a switch in mid-stream, though it must alter the center of gravity of a work, does not necessarily destroy its artistic integrity. On the contrary, in the particular case of *Evgenij Onegin* — where some sort of change undeniably occurs, intentional or not, and whatever the motive may have been — the change may legitimately be regarded as vital to the *fundamental theme* of the poem as it appears in its final form. For here too, as in the case of *Don Juan*, what "begins in fun ends in bitterness and sadness".[28] The carefree sparkle of Byron's first canto, already noted, is paralleled by an equal sparkle, an equal gaiety in the first chapter of *Evgenij Onegin*. But like Byron, Puškin is unable or unwilling to maintain his carefree gaiety. And this replacement of carefree gaiety by a more somber mood is central to the *fundamental theme* of *Evgenij Onegin*. The descent from poetry to prose; the conflict between idealism and disillusionment; the changing of the seasons; the passing of youth; the sorrow of time wasted never to be retrieved; the failure as the years go by to find a meaning: these are some of the component parts which produce the *motivo fondamentale* of *Evgenij Onegin*. And however

[25] *Ibid.*

[26] *Ibid.*, for discussion of this viewpoint.

[27] B. V. Tomaševskij, *Puškin* (Izd. Ak. Nauk SSSR, 1956), I, 613-614.

[28] Cf. footnote 16.

long and loud we talk of Puškin's advance in *Evgenij Onegin* from subjectivity to objectivity, it is not primarily Onegin nor Lenskij nor even Tatiana who most deeply arouses the reader's sympathies: the *fundamental theme*, I believe, is in the final analysis the feeling of the author, the *lyric hero* — Puškin's feeling of sadness, isolation and no-return.

Remembering that the *fundamental theme* of every work is unique, *is* in fact the work, it is nevertheless difficult to avoid the conclusion that the fundamental theme of *Evgenij Onegin* is — in spite of the many differences — not so very far removed from that of *Don Juan*. Both Byron and Puškin were in their own ways smitten by the Romantic sickness and for both men, however much they objectivized and intellectualized their sufferings, the sickness appears to have been mortal.

sufficient here to point to only one fact in order to explode the concept of some kind of distinct influence of the Montenegrin folk song on the formation of the verse in "Gorski vijenac". Having in mind the quantitative verse-ending tied to the use of bisyllabic words at the end of the line, I found, out of 1,352 lines of the folk type on the pages of Njegoš's collection "Ogledalo srpsko",[5] a total of 34 lines with a short first syllable in a bisyllabic word at the end.[6] Of these, there were 10 lines of the type:

... kako oni sada Srblje *kòrē*,

in which all words in the entire line, or in the second hemistich, were bisyllabic with short syllables. There were 11 lines of the type:

(od izdaje Brankovića Vuka)
njemu zato bila vječna *mȕka* ...

in which, at the end of the line, there is a bisyllabic word with a short first syllable because of rhyme. The remaining 13 lines were of the type:

da se ovo nadaleko *čȕjē* (five lines),

or the type:

... ni golema u Primorju *mòmka* (five lines),

or the type:

pa opeta evo vojsku *kȕpī* (three lines).

As can be seen, in the realization of the quantitative ending about which we speak, the correspondence between Vuk's folk *deseterac* and that of Njegoš in "Ogledalo" cannot be questioned, and furthermore, it can be seen that Njegoš's *deseterac* here is far below the poetic strength of the classical folk verse. Explanations can easily be found for the greater part of the deviation from the main tendency realized in Njegoš's "Ogledalo". First, one must exclude the 11 decasyllabics where the rhyme between the lines was the motive for placing bisyllabic words with short syllables. Rhyme between the lines is quite unusual in the *deseterac* of "Gorski vijenac". Furthermore, when we keep in mind that, in Vuk's folk poetry, lines with a series of bisyllabic words in the whole line

[5] See Petar Petrović Njegoš, "Ogledalo srpsko", *Cjelokupna djela*, V (Beograd, 1951), 189-237. The songs which Njegoš edited are about Montenegrin battles, and are sung in Montenegro.

[6] In percentages, 2.6%. In 34 verses of this type there are 292 lines with a bisyllabic word in which the first syllable is long, in percentages, around 21.6% (or 8ȁ times more lines which correspond to the tendency of Vuk's *deseterac*).

(or in the second part of the line) with only short first syllables are not unusual,[7] then we still have to explain, in particular, the entirely insignificant percentage of departures from the usual norm of folk poetry. If, with a good ear for the gusle-accompanied folk *deseterac*, we subject to analysis only two or three of the lines which I separated in this last group of lines, then we shall find a reliable explanation for both these and similar decasyllabic verses in Vuk's publications of folk songs. If we were to change the order of words in the line: "... da se ovo *nadalèko čȕjē*", in order to satisfy the quantitative verse-ending: "da se ovo *čȕjē nadalèko*", we would get a line somewhat more suitable to the ear which from childhood has accustomed itself to hear singing with gusle accompaniment. But if we were to change the line: "... pa opeta evo *vôjsku kȕpī*" to: "... pa opeta evo *kȕpī vôjsku*", we would not achieve anything that would tend to make the line sound more harmonious to us in terms of the spirit of the epic *deseterac*. And to an even lesser extent would we achieve anything if we were to give to the line: "da to bude *i prìncipu jȁdno*", the following form: "da to bude *jȁdno i prìncipu*". This last line, constructed thus, sounds almost like prose.

What does this mean, and whence comes the phenomenon that lines which conform more to one of the most dominant tendencies in the formulation of the folk *deseterac*, sometimes sound worse to us than those which stand in contrast to that same tendency? One can answer this question only when one enters into a more comprehensive examination of the means by which the folk singer effects certain poetic tendencies when their features are found to be in conflict with each other. I shall put aside for the moment an attempt to explain the examples given above in terms of contradictory tendencies. Before doing this, it is important to bring forth the first conclusion from an analysis of the examples taken from folk verses given in Njegoš's edition.

The folk verses of Njegoš's region were not able to influence the structure of his verse in "Gorski vijenac" because no difference between the Montenegrin and the classical folk *deseterac* (the one which Maretić investigated in the books Vuk published in his lifetime) really exists in the sense in which the *deseterac* of "Gorski vijenac" differs from the folk *deseterac*.

This can be confirmed by a mass of data which are not difficult to verify. Among the poems which collectors sent to Vuk from Montenegro and which were published only after his death, we can easily distinguish those which are purely of a folk type and have unadulterated folk coloring. And between these and the classical songs which Vuk noted down from the best singers, we can see no difference when we examine them in the light of the use of the quantitative

[7] In Njegoš's "Ogledalo", these types of verses exist in proportionally somewhat greater numbers, and that is because, otherwise, his folk *deseterac* would be much more monotonous then the *deseterac* in Vuk's collections.

verse-ending in the second part of the decasyllabic line. This can be confirmed in two poems from that region which Njegoš is known to have sent to Vuk.[8]

When we speak about the peculiarities of the *deseterac* as it is sung in Montenegro, and when we have in mind its structure alone, then we can characterize it reliably only when we carry out its classification on the basis of the following criteria:

1. One part of the Montenegrin folk songs comes from the best singers, and when they were written down by people who did not alter the characteristics of the folk verse in them they have the same qualities, in regard to the main metrical features, which we know from the classical verses of Vuk's folk songs;

2. One part of the Montenegrin folk songs comes from inferior singers and, as such, abounds in weak lines (for an examination of the structure of the folk *deseterac* they can be interesting only to the same extent as songs from other regions with similar characteristics, which we can find by the hundreds);

3. One part of Montenegrin folk songs comes from the pens of educated men who themselves either created them or, in their editing of folk songs, touched them up so much that they took away much of the true folk expression (in them the true qualities of folk verse can more or less be reflected, but data we take from them will always be misleading if we look at it from a purely statistical standpoint).

When we speak only about Njegoš's *deseterac*, then we must distinguish between the following categories of verse:

1. the folk *deseterac* merely written down by Njegoš (as it is in certain songs which he sent to Vuk);

2. the *deseterac* in poems written by Njegoš himself, where one must distinguish between two kinds of poems:

    a. Those which, in terms of the structure of the verse, are in general conformity with the classical folk *deseterac* and,

    b. those in which Njegoš had inferior poetic inspiration, so that one finds an abundance of verses which are nearer to prosaic than to poetic expression (and in both varieties of these verses one can see, here and there, the influences of literary expression — see above the occasional rhyming in songs from "Ogledalo," etc.);

3. the *deseterac* in "Luča Mikrokozma", which is set entirely apart by its particular structure (which is interesting to examine separately, but not when one is seeking folk characteristics in the *deseterac*);

4. the *deseterac* in "Gorski vijenac" and "Lažni Šćepan Mali", in which is shown Njegoš's true creative strength, not only in poetic qualities which are so obvious to us (and easy to perceive) in a study of his works, but also in the

[8] See Vuk, *Srpske narodne pjesme*, IV (1932), xxviii and elsewhere.

manner in which he rhythmically constructed his verse, a subject which has until now remained unstudied and almost unknown to us (although we can easily feel the rhythm of this verse as soon as the *deseterac* of "Gorski vijenac" starts to ring in our ears, no matter what section of this work is read).

The difference in the use of the quantitative verse-ending in the ninth syllable, as it is used in the folk *deseterac* and in the *deseterac* of "Gorki vijenac," needs to be illuminated more clearly than it has been up to now. First of all, the ninth syllable is taken into consideration only if there is a bisyllabic accentual whole at the end.[9] Lines of the following type quite conveniently correspond to the way in which Njegoš constructs his *deseterac* in "Gorski vijenac": "su dva mača i su dvije *krȕne*"; "Srpkinja ga još radjala *nìje* / od Koseva a ni prijed *njȅga*"; "ono nije bez nekakve *mȕkē*"; and so forth. However, we can easily determine that it is not irrelevant to Njegoš whether or not he places a bisyllabic word with a short first syllable at the end of the line. In the name *Crna Gora*, Njegoš changes the order of the words only when they are used at the end of a *deseterac*: "razvedri mi iznad *Gòrē Cȓnē*"; "kakvo roblje sasred *Gòrē Cȓnē*" (compare: "*Cȓnū Gȍru* pokorit ne mogu, ma nikako da je sasvim moja").[10] Why did Njegoš need to change the word order of a geographical name at the end of a line? The basic impulse here is just as it is in the folk *deseterac* — to realize a tendency towards the quantitative verse-ending: a bisyllabic word with a long first syllable at the end of the line. This can clearly be felt when we change this word order: "razvedri mi iznad *Cȓnē Gòrē*" (now this is not Njegoš's type of line, any more than it is the type of the folk *deseterac*). Njegoš does not willfully change the usual spoken order of words if he has no special reason for it: "Slušaj Vuče i *òstāla brȁćo*" — the spoken order remains although, for the quantitative verse-ending, it would be necessary to say: "i *brȁćo òstālā*". The same applies to "je li instinkt al' *duhōvnī* vȍdja". However, the question must be asked: Why was this word order, which is contrary to normal spoken word

[9] Here one must consider the fact that there has taken root the erroneous view that the shortness of the ninth syllable in the folk *deseterac* is relevant not only when it is at the end of a bisyllabic word, but also when there are three and four-syllable words in the same position. Corović also thought this, and various authors have repeated the same opinion: R. Jakobson, "Über den Versbau der serbokroatischen Volksepen", *Extrait des Archives Néerlandaises de Phonétique expérimentale*, III-IV (1933), 7; Kiril Taranovski, "Principi srpskohrvatske versifikacije", *Prilozi za književnost, jezik, istoriju i folklor*, XX, No. 1-2 (1954), 24; M. Franičević, "Kratak historijski pregled literature o nekim problemima našeg ritma", *Rad jugosl. akademije znanosti i umj.*, CCCXIII, 49, etc. In fact, it is not true anywhere in the folk *deseterac* that the long ninth syllable is avoided if some other word than a bisyllabic one stands at the end. The lines: "Pa ga eto u polje širòko ..."; "To se čudo čulo nadalèko ..."; "Udijeli kljastu i slijèpu"; and others fully correspond to the manner in which the *deseterac* in folk poetry is constructed. See also Maretić, *op. cit.*, p. 129.

[10] I mention the literary *štokavski* accent because, in the matter of quantity, it is all the same whether we do or do not take into consideration Njegoš's Montenegrin accent.

order, necessary to Njegoš: "Vuk na ovcu svòje právo ìmā" (and not: "Vuk na ovcu *ìmā* svòje *právo*", by which the desired quantitative verse-ending would be realized)? Or, why did it better suit Njegoš, contrary to the tendency toward the realization of a quantitative verse-ending, which, as we have seen is not irrelevant even to him, to use the word order in this line: "Ko na brdo, ak' i málō, *stòjī*, (više vidi no onaj pod brdom)"? In order to answer these and similar questions, it is necessary to take into consideration still another example of Njegoš's poetic phrasing in which the disordering of words which we find in "Gorski vijenac" would result in a changed line that would no longer have Njegoš's style and strength of expression, and would sound more like prose than poetry. If, for example, in the line: "Ja ću počet kada drȕgī néćē", we change the form so that it reads: "Ja ću počet kada *néćē drȕgī*", the form would be nearer to prose than to the real strength of poetic expression in the *deseterac*, both folk and artistic. We shall feel this even more if, in a similar way, we change the form of the *deseterac* in the lines: "slabost'ma smo *zȅmlji prìvezāni*"; "Za vrsnijem brȁtom ili sînom / pusti glasi *mȉlōst* ustòstručē"; "iza plača *vesèlijē pȍjēš*"; etc., etc.

Here one of the main roles is played by sentence stress, that is, Njegoš's exceptional sense, possessed only by him, of the ordering of strong, stronger and weaker places in sentence intonation. No matter which of the lines given here we take (and no matter which lines we take from "Gorski vijenac" where, in this manner, lines of the *deseterac* with short syllables of bisyllabic or trisyllabic words at the end are interwoven), they will clearly confirm this. In the line "iza plača veselije pȍjēš", there are accentual wholes based on strength: the one with the sign ______ has the main accent: the one with ______ has the secondary strength; and the one with ______ is weak in strength of sentence intonation. Take, for example, the lines: "Vuk na ovcu svoje pravo ima"; "Ko na brdo, ak' i malo, stoji, / više vidi no onaj pod brdom"; "Jâ ću počet kada drugi neće"; etc., etc. (______ designates normal, entirely ordinary pronunciation in sentence intonation).[11]

If a bisyllabic word is at the end of the line, then it usually represents a weak (or a weaker) place in Njegoš's sentence intonation. Examples which would not confirm this rule are very rare, and there must exist other good reasons in the combinations of rhythmic features in order for such an instance to be justified. Let us take for example: "Junaku se često puta hoće / vedro nebo nasmijat grohotom". Words with a strong accent at the beginning and at the end of the line occurred for special reasons: the symmetry of sentence intonation which

[11] About the grading of accents according to strength in the accentual wholes and in sentence intonation see Br. Miletić, *Osnovi fonetike srpskoga jezika* (Beograd, 1952), pp. 71-97.

Njegoš favored was realized (the main sentence stresses fall on the beginning and the end of the line). The dominant tendency to which his rhythm is subjected can be described as follows: the bisyllabic accentual whole at the end does not seem to be the single carrier of the main stress in the line. Speaking generally, the end of the line in Njegoš's *deseterac* is usually a weak place in sentence intonation if, after the caesura, it consists of more than two accentual wholes. A more strongly accentuated unit can occur there, with rare exceptions, only if it is trisyllabic (compare: "više vidi no onaj pod brdom"), but in this example there is no dominant tendency toward main sentence stress.

The way in which Njegoš's *deseterac* in "Gorski vijenac" (and, to a certain extent, in "Šćepan Mali" too) differs from all other kinds of Serbo-Croatian *deseterac* is in his special talent for the combining of syntactic wholes and the arranging of accentual wholes into variable patterns of sentence intonation. Besides the fact that a bisyllabic word at the end of the line usually occupies a weak place in the sentence intonation (which represents one of his most prominent rhythmic tendencies), Njegoš also has rhythmic features which are realized to a greater or lesser degree according to necessity in the various clashes of rhythmic tendencies (accentual, accentual-quantitative, and syntactic) so that one tendency or another may predominate.[12]

Fr. Wollman was on the right track in his study of Njegoš's verse when he attempted to discover the main characteristics of the *deseterac* in "Gorski vijenac" in Njegoš's arrangement of syntactic stresses. Methodologically Wollman found the most reliable basis for the study of the structure of Njegoš's verse.[13] But he made one great error by using for his analysis of syntactic stress the recitative pronunciation of people from Njegoš's region as it was traditionally adopted in a special type of declamation. In order to interpret accurately Njegoš's verse in relation to the spoken intonation of the sentence, and in order to intonate properly the semantic wholes (the changing into poetic forms of a characteristically strong and rhythmically dynamic expression), it is necessary to take as a starting point the intonation of ordinary speech as it appears in the poet's region (in all variations of emotional and rational-emotional means of expression). Wollman did not have this essential foundation

[12] About the clash of rhythmic tendencies as a general phenomenon in the structure of verse, see Jakobson, "Über den Versbau ...," p. 7.

[13] Before Wollman, Ćorović pointed to the peculiarities of sentence intonation in Njegoš's verse (see *Misao*, XIX, 1374), but he was far from interpreting Njegoš's sentence stresses correctly. Entirely counter to what we would have expected from a man who feels the peculiarities in the syntactic accents of the verse in "Gorski vijenac", Ćorović sought influence on Njegoš's verse in this work in the *deseterac* of Sima Milutinović. However, even the most extensive analysis can show how far from each other in their rhythmic qualities are Njegoš's verse in "Gorski vijenac" and Milutinović's verse in "Srbijanka", or in the drama "Dika crnogorska". (Sometime I shall demonstrate this in more detail.)

when he undertook an analysis of Njegoš's *deseterac*. And a random sampling of Wollman's analysis attests to this fact. Only in the declamations of the type which Wollman heard can the sentence stress of Njegoš's verse be expressed: "jedna slamka među vihorove", as Wollman wrote it down (*Slavia*, IX, 759). First of all, the stress "jèdna [Montenegrin pronunciation is jȅdna] slâmka *mȅđu* vȉhorove" can be the stress in declamation, but never the stress of ordinary speech ("među‿vȉhorove" and not "mȅđu vȉhorove").[14] The Montenegrin form of recitation is a kind of singing, and therefore each stress it has is a singing stress, which only to a certain degree can conform to spoken stress, and we must never seek in it standards for the determination of syntactic stress.[15] The main stress in syntactic wholes arranged the way they are in this line is not at the end (as Wollman thought), but rather at the beginning of the line: "Jedna slamka među vihorove". In the line: "sȕza mòja nêmā ròditelja", the strongest stress occurs after the caesura (-/nêmā), and the weakest at the beginning (suza moja). Whoever has a feeling for Njegoš's stress in its hierarchy of intensity will determine that, in Njegoš's line: "Zloga gosta Evropi Orkana!" the strong, rather the strongest, place is at the beginning (Zlôga gȍsta Evrópi Orkána"), but Wollman, upon hearing Montenegrin declamation, felt it to be at the end of the line.[16]

In the arrangement of syntactic stresses there exist basic correspondences in the folk *deseterac* and in the *deseterac* of Njegoš's "Gorski vijenac", but when one goes a bit beyond these basic tendencies, then great differences become apparent. What, in the folk *deseterac*, appears as one of the more commonly manifested tendencies appears, proportionally, more rarely in Njegoš's work; in precisely this manner he in his own way avoids the monotony that might otherwise result. The basic tendency of the syntactic accent in the folk *deseterac* shows itself in lines which are repeated, such as the beginnings of the epic folk

[14] Students of our *deseterac*, both domestic and foreign, make glaring errors in their delimitation of accentual wholes in verse. Thus, for example, the line from the folk song "Ōj Mȉlošu, naš ròdjenī brȁte" (this is the proper accentual notation) is accented: "*Oj*, Mȉlošu, *nȁš* ròdjenī brȁte" (K. Taranovski, "O jednosložnim rečima u srpskom stihu", *Naš jezik*, nova serija, II [Beograd, 1956], 28), which sounds completely affected in both conversation and recitative intonation of expression.

[15] Even if he, as a foreigner, had not erroneously started from declamation, Wollman was able only with difficulty to determine correctly Njegoš's syntactic accents. In many cases, it is not easy even for people who feel Njegoš's *deseterac* most intimately to grade his accents according to the strength of sentence intonation. As in the case of other writers, Njegoš's verse, usually constructed in concentrated poetic expression, demands that its interpreter try to uncover the poetic-philosophic thought, and very often the strength of the accent will be determined by that thought which is most suited to accompany the sense of Njegoš's words in all their nuances.

[16] I leave for another time the task of showing in more detail the errors in Wollman's analysis of Njegoš's sentence intonation.

songs, or which are otherwise repeated in the songs. For example: "*Konje jašu do dva pobratima ...*"; "*Vino piju do dva pobratima ...*"; "*Vino pije Kraljeviću Marko ...*"; "*Nadaleko zaprosi đevojku*", and so forth. The tendency that is expressed here to a certain degree, but is manifested in a large number of cases in the epic folk *deseterac*, can be characterized by several points:

a. if there are four accentual wholes, two and two in segments separated by a caesura, then the stronger accents are located in the middle of the line before and after the caesura, and as a rule the one that is in the second segment of the line is stronger: "Kònje jȁšu / dò dvā pòbratima";

b. if there are three accentual wholes, then the main accent will again be at the same place, after the caesura: "sve‿glȅdaste / dànas nè‿viđoste // đe‿pródoše / Lȉjevanjāni Tûrci";

c. if there are four accentual wholes in the line arranged so that one is before and three are after the caesura, then, preferably, the main stress comes immediately after the caesura, and the remaining ones will be freely arranged according to strength: "Đèvōjka‿je / svòje ȍči kléla";

d. if there is a three-syllable or four-syllable accentual whole at the end of the line, then it can be the bearer of the main sentence stress, and it often happens that way: "Kad‿se‿žȅnī / sȓpskī car-Stjȅpāne ... Kad‿se‿šćáše / po‿zèmlji Sȓbiji, po‿Srbiji zèmlji da‿prèvr̄ne ...."[17] With a few more examples, compared in this way, a point of departure might be found by which the possibilities for a variety of intonational patterns in sentence stress could be shown in greater detail (but this will be discussed another time).

One important phenomenon needs to be kept in mind: the main point of sentence intonation is to be found somewhere in the second segment of the *deseterac*. Preferably this point is located immediately after the caesura. But since atonic words (proclitics or other kinds of words which, according to the nature of the language, can be optionally atonic) often appear in this position, the main stress is often shifted toward the end of the line. It is possible to find

[17] It is necessary to reject still another erroneous conception about the nature of the folk *deseterac*, which has to some extent taken root in scholarly literature, but which does not correspond to reality, and this is the statement that among the characteristics of the folk *deseterac* is the rhythmic tendency for the *deseterac* to be realized in five bisyllabic accentual wholes (see, among others, R. Jakobson, "André Vaillant: Les chants épiques des Slaves du Sud...", *Byzantinoslavica*, IV, No. 1 [1932], 194). Because of this, it seems to me, our *deseterac* is incorrectly treated as verse with trochaic "phrasing", instead of its value being more correctly defined in terms of trochaic-dactyllic verse. If there are no separate properties in the sense of vocalic sound effects (rhyme, assonance, alliteration, etc., for example: "...svȕd su brȍdi đe gȍd dôđēš vȍdi", and others), then nearly every *deseterac* with five accentual units sounds monotonous. Consequently, I am not taking this kind of structure into consideration; it does not correspond to that ideal scheme of verse which soars in the consciousness of the creator of the folk epic when he forms the classical epic line.

an entire song, or an entire series of songs, in which there is not a single main sentence stress in the first part of the *deseterac*. But in songs of a more dynamic style, the main stress will turn up at the beginning of the line or in general somewhere in the first part of the *deteserac*. And in this way the singer creates a moment of "deceived expectation". This does not happen very frequently in the folk *deseterac*, but when it does occur, then we feel an extra strong rhythmic effect. In the verses of Vuk's greatest storytellers we can notice this phenomenon more frequently than anywhere else in folk songs. Starac Milija clearly felt the force of this effect; therefore he knew how to begin a song in a way in which it ordinarily does not begin: "Netko bješe / Strahinjiću bane", and at the end he again, in a similar way, would intonate the line: "Pòmalo‿je / tàkvijeh junákā kō‿što‿bjȅše / Stràhinjiću bane", so that his entire sentence in a double line would acquire a symmetrical shape, with the main stresses at the beginning and near the end. After three beginning lines with the sentence stress in the second segment of the *deseterac*, Filip Višnjić, in the song "Početak bune protiv dahija", creates a moment of "deceived expectation" in such a way that, in the fourth line: "i‿da‿drȕgā / postane sudija", the main stress crosses over to a position before the caesura (and, in so doing, breaks the syntactic whole "druga sudija" in the sense of symmetric evenness). We can compare the following arrangement of syntactic stresses: "Ako mèni do nèvoljē dôdjē, jā‿ću‿làsno kȍplje okrénuti". The folk singer can achieve an exceptional rhythmic effect when he arranges the stresses from the beginning to the end of the line so that there is a successive decline in strength: "Krîvo bȉlo Golòtrbu Ívu", and after that: "đe‿se‿njèmu smȉju kapetáni", and so forth.

Without studying the syntactic features and without taking note of the syntactic verse-endings achieved in keeping with the rhythmic tendencies of the line, it is impossible to determine more fully and more comprehensively the characteristics of either the folk or of Njegoš's *deseterac* (or, in general, of any kind of verse). Only when the syntactic verse-endings are more exhaustively examined can we more fully understand aspects of phenomena which are pointed out to us in the studies of the Serbo-Croatian *deseterac* done up to the present.

In order to evaluate more easily the difference between Njegoš's and the folk *deseterac* in the matter of sentence intonation (and here, in fact, is where the main differences are reflected) after all that we have been able to observe in the structure of the folk *deseterac*, it would be best to begin with an analysis of several lines that clearly and forcibly express Njegoš's dynamics. Let us listen to the sentence intonation in the following lines:

Bješe oblak sûnce uhvatio

Bješe tama gȍru pritisnula,
Pred oltarom plȁkāše kandjelo,
Na gusle se strȕne pokidale,
Sakrile se vîle u peštere ...

Several basic features of Njegoš's sentence intonation are contained in these few lines which, from the very first, give us the impression of having rhythmic features entirely different from those that are found in the folk *deseterac.*

1. Lines of this type by Njegoš, when examined both in isolation and in long series of decasyllabic wholes, have as their principal feature a *super-saturation* of sentence intonation. This results from the basic general features of his style: the concision, the compactness of his poetic expression. Rich in the possibilities for expressing idea and feeling in poetic picture, in metaphor, allusion, and allegory, with a strong potential in meanings contained in the sentence wholes, Njegoš's work is to the same degree rich in possibilities for evoking a poetic picture with his own distinctive combination of all tonal effects, and his own distinctively formed line of sentence rhythm and sentence melody. In every place in Njegoš's "Gorski vijenac" where the range of feelings and of contemplative activities rises above the level of ordinary speech, where the contemplative and emotional force of the poet must reach a climax of the most moving expression, the same kind of saturation of sentence intonation which we have in these (or similar) lines will immediately be demonstrated. We can feel this in a slightly different way not only in Njegoš's "Kolo" [part of "Gorski vijenac", see line 198], when it says:

Bôg se drâgī nà Srbe rȁzljūtī
za njȋhova smȑtnā sagrešénja, etc.,

but also in the conversations when the lines, for one reason or another, acquire an atmosphere of greater excitement:

Mȑkī Vûče, pòdigni bȑkove
da ti vȉdju tȍke na pȑsima,
da prȅbrojīm zȑna od pȕšākāh
kolika ti toke izlòmiše!
Mrtvu glavu ne diže iz groba. ...[18]

When it was necessary for Njegoš to present, in the words of Prince Janka, the

[18] It would be necessary to present an analysis of the strength of sentence accent in greater detail than I have been able to do here. The main point is that these matters be grasped in principle.

clash in the conversation between Vuk Mićunović and Hamza Kapetan, then, at the point of transition from the more reserved dynamics of ordinary narration ("Ka iz rúgē tô bī iz počétka ...") after the line: "uljegoše u krupne riječi", he had to describe the "momentous words" ("krupne riječi") of the conversation with the same saturated sentence intonation found in the style of the lines:

Raspali se Mićunović Vuče,
pa se Hamzi poprimače blizu ...
đe izdajnik bolji od viteza ...

(followed by a whole series of similar lines down to the end of this narrative). On another page, the large number of dialogues having the tone of ordinary speech (questions and answers) demonstrates a less extreme oscillation in the level of sentence intonation. The entire conversation of Duke Draško with his relatives is given on this other, more reserved level of sentence intonation. The alternate arrangement of conversational sections in "Gorski vijenac" which we might be able to indicate by the terms *saturation* (expressed in varying degrees) and *lack of saturation* (the usual conversational dynamics of expression), is the quality of sentence intonation which sets apart Njegoš's verse from all the creative poetic work that has been written in Serbo-Croatian.

How Njegoš transposes the folk *deseterac* into his own rhythmic pattern in those places where it serves as the impulse for the formation of his verse in "Gorski vijenac" can best be demonstrated by this example. Having carried over into his verse the folk line: "sad s' u mene raspale plamovi ...", he gave it his own expressive trappings: "raspale me užasa plamovi".[19] Here we simply are unable to determine where the main stress is; each accentual whole in the line is given a full measure of poetic accentuation to such an extent that the expression is saturated with sentence intonation. *In the possibilities for the saturation of sentence intonation we have, then, one of the principal marks by which we can distinguish Njegoš's* deseterac *from the folk* deseterac.

2. We shall return once again to the verses which I took as a main point of departure in the examination of rhythmic characteristics of the sentence in "Gorski vijenac" in order to discover other peculiarities of Njegoš's rhythm (pp. 202-203). In four lines the main stresses fall immediately after the caesura (a tendency which also exists in the folk *deseterac*), and the arrangement of primarily and secondarily strengthened stress is given in various forms of intonational symmetry: (a) in the first line the second stress, in terms of strength, stands at the beginning of the line (the two sentence stresses are nearly equal in strength, and (b) in the second and third lines, stresses of secondary degree are

[19] See Pavle Popović, *O Gorskom vijencu*, (Beograd, 1923), p. 189.

located in accentual wholes at the beginning and at the end of the line (see the above marks). Symmetry is achieved doubly within the boundaries of lines taken as wholes and within the boundaries of a series of lines as they follow one another. Again, this is one of the characteristics which are to be found only in Njegoš's *deseterac*. This form of symmetrically combined rhythmic line, which is achieved not as an isolated phenomenon but in a great number of lines, will not be found in folk poetry at all and only rarely anywhere else. In this and similar arrangements of sentence stresses is contained that propèrty of Njegoš's verse which we feel from the first rhythmic impression the verses make upon our ears:

Čȁšu mȅda jošt nȉko nȅ popī
dok je čȁšōm žȕči nȅ zagȑčī.
Čȁša mȅda ȉstē čȁšu žȕči,
Smȉješāne nȁjlakše se pȉjū.

Folk decasyllabics cannot be ordered into units like these. They cannot have this kind of variability in intonational symmetry that would produce a general rhythmic effect in the form of one special composition.[20]

Symmetrical evenness in the arrangement of accentual wholes within the line, and *variable alternation in the succession of symmetrically arranged sentence stresses* are still another of the main characteristics of the *deseterac* in "Gorski vijenac."

3. A person with a sharp ear can notice still another characteristic of Njegoš's rhythm in the lines already presented: "Bjȅše táma gȍru prȉtisnula ..." (Njegoš's stress is "tâma" and "pritȉsnula") and in those immediately above: "Čȁšu mȅda jošt nȉko nȅ popī. ..." In the first group of lines, all the words immediately after the caesura which, as we saw, carry the main sentence stress, also have falling accents at the beginning. In the second group of lines, all the stresses are at the beginning of the word; that means they are falling to both Njegoš's ears and our ears. This is not an accidental phenomenon. Njegoš really felt the power of stress at the beginning of the word which was the bearer of a strong sentence stress. A more detailed analysis will show the extent to which this is true. That both the quality of stress (and not only the quantity) and its place in the accentual whole play a definite role in Serbo-Croatian *deseterac* can be established if one examines other decasyllabic verse besides Njegoš's. Even for the classical folk *deseterac* it is not entirely irrelevant whether a falling or rising accent will come at the place where there is a tendency to form the main sentence stress, or whether the place will be atonal. One analysis of mine might be interesting in

[20] The symmetry in Njegoš's *deseterac* must be regarded as a tendency. In these verses it is partially disturbed, but its features are shown in them.

this connection. In three folk songs which Vuk wrote down from his three best storytellers, Filip Višnjić, Tešan Podrugović and Starac Milija, he himself wrote down the stresses in the accentual units after the caesura (where, as we have seen, the main stress in the line usually falls). I took into account the possible differences in accent which can vary according to the region from which the singer comes ("tȅbe" or "tèbe" and the like). In the 628 lines of Visnjić's song "Početak bune na dahije", after the caesura there occur 243 falling accents, 210 rising accents, and 184 unaccented syllables. In Podrugović's song "Ženidba cara Dušana", we find for 697 lines the relationship: 349:198:150. In the song "O Strahiniću Banu" (Starac Milija), the relationship is 362:243:205. The general picture is sufficiently clear: in all three songs the falling accent dominates in the first syllable after the caesura. In "Ženidba cara Dušana" there are as many falling accents in this position as there are syllables with rising accent and with no accent put together. Of course, this proportion does not correspond to the relationship of accents as they appear in the spoken language. In it still another rhythmic tendency can be discovered which creates a qualitative difference between the accents in prescribed places in the line (about which it would also be worthwhile to speak in connection with the character of the trochaic line in folk epic phrasing). Here it was important for me to compare the line of "Gorski vijenac" with the classical folk line, and to see how each of them appears in the light of this separate rhythmic tendency, which is of a purely accentual character (considering the quality of the accent at a prescribed place in the line). And this very type of comparison enabled us to confirm the impression which we were able to draw from an analysis of Njegoš's verses quoted above. In 810 lines of "Gorski vijenac", counted from the beginning, I established the relationship of 456 (falling accents at the beginning of the word immediately after the caesura): 152 (to Njegoš, falling accents after the initial syllable in an accentual whole, and to our ear, rising accents on the initial syllable); 222 (with the accent further from the initial syllable). Of course, here I also included in my calculations Njegoš's falling accents at the beginning of words to which correspond the rising accents of the newer type in the same place ("vȋla": "víla", "nȍga": "nòga", and so forth). But, on the other hand, the proportionately much greater number of unaccented syllables (by the standards of both the new and the old, i.e. Njegoš's, accentual system) can be interpreted only by Njegoš's more frequent use of atonic words right at the beginning of the second segment of the line than is the case in the folk *deseterac*. (This peculiarity arises by itself from the conversational situation and the dialogue form in which Njegoš's work is given).

In any case, it remains a clear fact that one of the poetic rhythmic features common to both the classical folk *deseterac* and to Njegoš's *deseterac* appears dominant, and that the *accent at the beginning of the word after the caesura in the* deseterac *of "Gorski vijenac" is, in some places, a distinctive feature of Njegoš's* rhythm, and that, by this very element, Njegoš's type of rhythmic dynamic is made complete.

This phenomenon, as I have indicated here, can in my opinion be doubly important: (a) as much for the perception of general characteristics in the structure of the *deseterac*, as for the perception of differences between the folk and Njegoš's *deseterac*, and (b) for the confirmation of my hypothesis that the place after the caesura is attractive for the main accent in both the folk and in Njegoš's *deseterac*.

More detailed analyses will reveal something more about matters to which, for now, one can only allude. These verses, written in a specially expressed emotive tone, can show that the accent of the initial syllable is very important for Njegoš when the accentual whole occupies a strong position in the sentence intonation:

rûke stȁre, u njih mȁč i kȍplje,
kȑvāve mu rûke i òrūžje,
kȍracima bròjī tȕrsko trûpje,
skâče stȁrac kako hȉtro mòmče ...

Again we have a series of lines with accents on the initial syllable in the accentual whole which (now in another position) carries the main sentence stress.[21]

As I have tried to show here, I believe there can be no doubt that both the folk singer and Njegoš, creating rhythmic features as constants (unchanging), as dominants and as more or less attainable tendencies (both in similar and in different ways), appreciated the great role of those tendencies in which the combinations of sentence stress are expressed as the verse is formulated. And there is even less doubt that one is bound to perceive the individuality of Njegoš's rhythm in the variety and in the uniquely expressed peculiarities of Njegoš's sentence intonation. And when all this is kept in mind, it is not difficult to put aside any thought which would lead to the conclusion that Njegoš had any other impulses for the construction of his verses besides those which the classical folk song was able to give him. Similarly, when we limit ourselves exclusively to the comparison of the epic folk *deseterac* and Njegoš's *deseterac*,

[21] I have not spoken here of still another quality which clearly characterizes the dynamics of Njegoš's *deseterac*, and which is responsible for the saturation of sentence accent (with its variable rhythm in the sense already examined): that is *pause* and *tempo* (dependent upon the more expressive pause). My future work in Serbo-Croatian verse will be carried on along the lines of intonational symmetry, and the intonation of the pause and of expressive tempo.

it will not be difficult to arrive at the distinguishing features by which one can clearly perceive the extent to which Njegoš was independent and inimitable in the creation of this type of verse.

(Translated from the Serbo-Croatian by Frederick B. Johnson, University of Toronto.)

CERVANTES' "GYPSY MAID" AND PUŠKIN'S "THE GYPSIES"*

JACK WEINER
University of Kansas
and
E. F. MEYERSON
Johns Hopkins University

Much has been written on the influences which the French and English writers had on the work of Puškin. Comparatively little has been published regarding the influences of Spanish literature on the Russian poet; yet it will be seen that he was acquainted with both the language and literature of Spain, particularly that of Cervantes, whose exemplary novel, "The Gypsy Maid", served the poet as a text for certain of his exercises in Spanish translation. This study proposes to indicate possible influences of that work on Puškin's verse narrative, "The Gypsies".

Puškin's interest in Cervantes and Spanish literature is clearly apparent in the countless observations and references found in his correspondence and literary commentaries, as well as in his poetry. This interest was not entirely incidental. In his own family, his great-grandfather, the Negro Hannibal, had fought in Spain (1719-21). The poet alludes to this in his incomplete novel on his ancestor's life, *The Negro of Peter the Great.*[1] Numerous references to Spain appear in the study Puškin made on the Russia of Peter the Great.[2] Nor was his interest merely superficial. On comparing the invasion of Spain by the Moors with the Tartar invasion of Russia, Puškin makes the very discerning comment that the Tartars bestowed upon Russia "neither algebra nor Aristotle".[3]

In his own literary output there are various works with Spanish themes and Spanish backgrounds. "The Stone Guest" is the traditional Spanish legend of Don Juan; there are also "A Spanish Serenade",[4] "Inesilla, I am Here",[5]

* This is a revised version of an article which originally appeared in Spanish in *Nueva Revista de Filología Hispánica*, XVII, (1963). The authors thank Mr. Antonio Alatorre, the editor, for permission to publish it here.

[1] A. S. Puškin, *Polnoe sobranie sočinenij* (Moscow-Leningrad, 1937-1959), VIII, 3.

[2] *Ibid.*, X, 18.

[3] *Ibid.*, XI, 513.

[4] "Ispanskij romans", *Polnoe sobranie*, II, 345.

[5] "Ja zdes', Inezil'ja", *Polnoe sobranie*, III, 239.

"Rodrigo", which he translated from English,[6] as well as the fragments of a novel in which the protagonist is a Spaniard.[7]

Being a political exile himself for over six years (May 1820-September 1826), Puškin viewed the Spain of his own era as a symbol of the struggle of liberty over tyranny; he read everything possible on the political situation in post-Napoleonic Spain. French intervention then threatened her democratic forces, and to Puškin's mind Colonel Rafael del Riego was the principal hero of the day. Riego and the Spanish struggle furnish the central themes of various of his poems, particularly "Skazali raz carju", written in 1825.[8]

P. V. Annenkov says in his biography: "Puškin learned English and Italian in the South [i.e., during his exile in Kišinev and Odessa in 1820-1824] and read much in both languages." Then he adds in a footnote: "The poet's father, Sergej L'vovič Puškin, notes that his son studied Spanish along with Italian."[9] What S. L. Puškin actually wrote was that his son "learned Spanish in his mature years".[10] There is no known evidence confirming Annenkov's assumption that S. L. Puškin's phrase "in his mature years" refers to the period of his son's exile in Kišinev and Odessa. Indeed, the evidence points rather to the period after Puškin's return to Mixajlovskoe. He himself wrote from Mixajlovskoe in September 1825 to his friend P. A. Katenin: "I do not know Spanish."[11] This statement is so remotely connected with its context that it can perhaps be best explained as psychological evidence of Puškin's growing interest in learning the Spanish language. The fact of his having studied it is confirmed by B. Modzalevskij's bibliographical description of his library, which lists half a dozen Spanish dictionaries and grammars, including one given to him by his friend S. A. Sobolevskij with the following inscription: "To Puškin from Sobolevskij in memory of the Rhine wine."[12] This inscription is undated, but it can scarcely refer to any other time than the period between the end of Puškin's six years of exile in September 1826, when he and Sobolevskij began frequently seeing each other in Moscow, and Sobolevskij's departure for Western Europe in January 1829.

The supposition that this was the period of Puškin's Spanish study is further confirmed by two translation exercises that he made from Cervantes' story

[6] "Rodrig", *Polnoe sobranie*, III, 445.
[7] "Gosti s"exalis' na daču ...", *Polnoe sobranie*, VIII, 35.
[8] *Polnoe sobranie*, II, 378.
[9] P. V. Annenkov, *Materialy dlja biografii A. S. Puškina*, A. S. Puškin, *Sočinenija*, I (Petersburg, 1855), p. 95.
[10] *Puškin v vospominanijax sovremennikov* (Leningrad, 1950), p. 21.
[11] *Polnoe sobranie*, XIII, 225.
[12] B. Modzalevskij, "Biblioteka A. S. Puškina (Bibliografičeskoe opisanie)", *Puškin i ego sovremenniki*, IX-X (Petersburg, 1910), Items 579, 827, 936, 937, 1036, and 1224.

"La gitanilla" ("The Gypsy Maid"), the later of which is in a notebook dating from 1828-1833.[13] According to Modzalevskij Puškin's library contained a four-volume set of Cervantes' *Novelas exemplares*, published in 1816, which was uncut except for pages 1-32 of Volume I. These pages contain the beginning of "La gitanilla", including the two passages Puškin used for his translation exercises into French and back into his own Spanish.[14] Why did Puškin select this particular story for his translation exercises? One possibility is that when he began studying Spanish he chose for his first reading and translation material a story with which he was already familiar. And how did he happen to be so familiar with "The Gypsy Maid"? We believe it is because he had carefully read the story (probably in French translation) before he sat down in 1824 to write his own narrative poem "Cygany" ("The Gypsies"), which reveals so many striking parallels with Cervantes' work that the possibility of direct influence cannot be dismissed.

The two stories have, in fact, the same theme — the theme of initiation and moral education, the theme of self-knowledge achieved through temporary reversion: the knowledge that civilized man cannot wholly discard either the old values or the social prejudices of the society which nurtured him.

Both Cervantes and Puškin present the idyllic and romantic in Gypsy life. The immediate plot of each story reveals a hero who has abandoned life in cultured society for love of a Gypsy girl; parallel with this leitmotif in both works runs the drama of jealousy and the moral responsibility of personal liberty. Both heroes mistakenly identify the sentiment of love with the laws of ownership, and to both heroines this principle of civilized society is altogether unknown. "Know that my freedom and I must go hand in hand",[15] Cervantes' Preciosa tells the love-stricken Andres; and to Puškin's Zemfira the idea of restraint is equally beyond comprehension. Both authors begin their narrative with a brief discourse on the ways of the Gypsies, and both tales close with the return of the hero to the society from which he came.

Each hero is welcomed into the group, initiated into the mysteries of Gypsy life, and accepted as a member by the male Elder of the tribe. "Be one of us",[16] Zemfira's aging father tells Aleko; while the old Spanish Gypsy bids Andres to "Rest now in the nest under our wings".[17] As the character of each hero unfolds, both Andres and Aleko are passionately jealous and deeply in love, but to neither heroine does this seem sufficient reason to give up everything and

[13] *Rukoju Puškina* (Moscow and Leningrad, 1935), pp. 83-87.
[14] Modzalevskij, *op. cit.*, p. 188.
[15] Miguel de Cervantes, *Novelas Ejemplares* (Barcelona: Ramón Sopena S. A., 1956), p. 43.
[16] *Polnoe sobranie*, IV, 180.
[17] *Novelas Ejemplares*, p. 64.

become a homeless Gypsy. "The are eyes which, bedazzled, mistake glitter for gold", the knowing Preciosa tells Andres, with a wisdom beyond her fifteen years, "yours may soon learn the difference".[18] Puškin's heroine is no less apprehensive: "Do you not regret all you have left behind forever, your cities, your countrymen?"[19] she inquires of Aleko.

Both heroes, however, are willing to renounce civilization without a backward glance; Aleko denouncing "the bondage of stuffy cities",[20] and Andres cheerfully abandoning "his profession of gentleman-at-arms and the vain glory of his illustrious lineage, regretting only that he had not learned sooner of that gay Gypsy life."[21]

But neither is able to complete his reversion. Andres is never able to discard his moral precepts or his status as a gentleman of society. For him to join the Gypsies in neighborhood thievery or cut the purse of a wayfarer are impossible tasks; and though he has been instructed by the Elder on Gypsy conduct toward the law and the importance of "not treading the empty air in the flower of our youth",[22] he falls a victim to his cultured upbringing in his very first encounter with the authorities. Called "a thieving Gypsy rogue", and struck by a soldier, the angry knowledge that he is a "gentleman" betrays him into avenging this slight to his personal honor. Running the soldier through with a sword, he brings near-tragedy to himself and the band.

Like Cervantes' hero, Puškin's has succeeded in becoming a Gypsy only externally. His revolt against society is little more than superficial; while he scoffs at the slavery of those who belong to the higher stratum of society, he remains himself a prisoner to their formal precepts. Aleko willingly accepts the freedom the Gypsies offer him, but his own inability to grant that freedom to others culminates in the closing tragedy. Bound by the moral precepts of the society he scorns, he cannot reconcile himself to the Gypsy girl's free and natural choice of another man when she has ceased to love him. Aleko kills both Zemfira and her young Gypsy lover, and the horrified nomads, who have no fondness for blood or vengeance, send him back to his civilized society.

Another characteristic common to both works is the notable absence of any Gypsy idiom. In Cervantes' case it is rather extraordinary, for his many travels about Spain brought him into contact with people of every class of life and manner of speech; his familiarity which their customs and language is faithfully

[18] *Ibid.*, p. 62.
[19] Puškin, *Polnoe sobranie*, IV, 185.
[20] *Ibid.*, IV, 185.
[21] *Novelas Ejemplares*, p. 61.
[22] *Ibid.*, p. 64.

set down in many other works. He unquestionably knew the Spanish Gypsies,[23] but for some reason "preferred not to inject Gypsy jargon into this work".[24] Puškin was also acquainted with Gypsy speech, but he too preferred not to use it in "The Gypsies".

Similar technical devices of "The Gypsy Maid" are also employed in "The Gypsies". Both heroes fear being apprehended, Andres by his father, and Aleko by the authorities; both authors introduce a secondary male character essential to the background of jealousy and passion. Though physical characteristics are seldom sharply drawn in romantic themes, the dark, caressing eyes of Preciosa match those of Zemfira; and if Preciosa is but fifteen, Zemfira "is but a child".[25] Each author reveals the temperament of his heroine in her song:

Preciosa:	*Zemfira*:
In this amorous enterprise In which love has me engaged I hold it greater happiness To be chaste than beautiful.[26]	Summer burns not so hot, Spring is not so bright as he; He is young, he is brave; What a love is his for me![27]

The contrast between Preciosa's chaste love and Zemfira's physical love reflects the contrasting viewpoints of the men who created them. Cervantes held that virtue was the highest moral value,[28] and spiritual love the most sublime concept of love. In women Cervantes sought the Ideal, and Preciosa is the image of freedom without physical abandon.[29] For Puškin, on the other hand, spiritual love is not love at all; and nearly all of his heroines surrender themselves physically to their lovers. Zemfira's ardent nature could not be otherwise; it reflects the typically Puškinesque concept of romantic love, particularly the concept of his early years. What has been said of Cervantes' heroine applies no less to Puškin's: "In the figure of Preciosa, Cervantes has drawn the ideal feminine image of a personality created in liberty and developed without hindrance in beauty of body and soul."[30]

For Puškin as for Cervantes, liberty was as necessary to the individual as the air he breathed; and in these two works the authors have embodied the deeper significance of personal liberty and the individual's moral responsibility to that freedom. If the key to the moral significance of "The Gypsies" lies in the wis-

[23] Agustín G. de Amezúa y Mayo, *Cervantes, Creador de la novela corta española* (Madrid Consejo Superior de Investigaciones Superiores, 1958), II, 5.
[24] *Ibid.*, II, 26.
[25] Puškin, *Polnoe sobranie*, IV, 193.
[26] *Novelas Ejemplares*, p. 83.
[27] Puškin, *Polnoe sobranie*, IV, 189.
[28] Amezúa y Mayo, I, 91.
[29] José Sánchez Rojas, *Las mujeres de Cervantes* (Barcelona: Montaner y Simón, 1916), p. 27.
[30] Amezúa y Mayo, II, 14.

dom of the tribal Elder,[31] it will also be found in the words of the old Spanish gypsy. Each writer's concept of liberty lies also in the words of these two characters:

Cervantes:
The fear of losing honor does not trouble us, nor the ambition to augment it keep us awake. We support no factions, nor do we arise at dawn to present petitions or to wait upon magnates, nor do we beg favors.[32]

Puskin:
Leave us, conceited man! We are wild and have no laws, but we do not torture, do not kill. ... You are not made for our savage ways; you seek freedom only for yourself.[33]

[31] Tomaševskij, I, 616.
[32] *Novelas Ejemplares*, p. 60.
[33] Puškin, *Polnoe sobranie*, IV, 201.

THE PARADOXES OF REJ'S BIOGRAPHY

WIKTOR WEINTRAUB

Harvard University

1. REJ'S LIFE SKETCH

Sixteenth-century Polish literature can boast rich and lively prose. Religious and political ferment, close contacts with the West, especially with Italy, the excellent economic situation of the gentry, all acted as mighty stimuli. And they could operate the more efficiently, since the country enjoyed great political and religious freedom, as well as relative peace.

In this abundant crop of religious and political tracts and treatises, translations of the Bible, volumes of history and booklets of popular fiction, the biography Mikołaj Rej published under the name of Andrzej Trzecieski (or, Trzycieski) in 1568, in the bulky volume of Rej's miscellaneous writings *The Mirror* (*Zwierciadło*), short as it is, occupies one of the front places.[1] True, it lacks the fine polish characteristic of most of the Renaissance prose based on Ciceronian models, but it more than makes up for this with its vigor, raciness, and humor. Its subject matter is one of the leading writers of the period, a man with vigorous, although somewhat coarse personality. It is amusing and it gives the reader the feeling of coming into close touch with the specific spiritual climate of the sixteenth-century Polish gentry. Much of the writing of the period was conventional. That was the high price paid for classical schooling. But our life sketch is very original. Its author handles his biographical material in a strikingly unusual, idiosyncratic manner.

Nothing at the very beginning announces the singularity of this essay. It

[1] There exist several good modern reprints. The first critical edition is to be found in the second volume of *Zwierciadło*, ed. Jan Czubek and Jan Łoś (Cracow, 1914). It was published with copious notes as a separate pamphlet by Henryk Gaertner as: Mikołaj Rej, *Życiorys własny* (= *Książeczki staropolskie*, No. 3, Warsaw, 1925). The two most recent editions are: Mikołaj Rej, *Żywot człowieka poczciwego*, ed. Julian Krzyżanowski (= *Biblioteka Narodowa*, Series I, No. 152, Wrocław, 1956), pp. 645-655; and Andrzej Trzecieski, *Pisma polskie*, ed. Stanisław Bąk (= *Biblioteka Pisarzów Polskich*, Series B, No. 9, Wrocław, 1961), where in Part I, pp. 221-227, we find the text in a modern transcription, and in Part II, pp. 77-82, a fototypic reproduction of the original edition of 1568. In this paper the sketch is quoted after the handy edition of Krzyżanowski.

starts in the usual way with a disquisition on the "ancient and noble" house of Rej. It tells us about the protoplast of the family (the Silesian Nanker), about the Rejs' coat of arms (Oksza), praises the Rejs as remarkably decent people, and singles out one ancestor who chose a military career and was killed by the Tartars. All this is the normal material that we would expect in a sixteenth-century biography of a man who was *bene natus*.

Neither do we depart from the standard treatment when we pass to our protagonist's father, Stanisław. We learn that, owing to the help of a relative who was Archbishop of Lvov, he moved eastward and settled in Ruthenia. Then we are told about his two marriages and are given details about his second wife, the mother of our Mikołaj Rej, which should prove from how noble and rich a family she came.

Only after these habitual preliminaries do we proceed to the actual protagonist. And here, at once, we have a surprise. The biography proper starts with the story of young Mikołaj's piteous education, which is told with obvious gusto. Thus, we learn that Mikołaj's father, "a pious, honorable and quiet man", being little interested in public affairs did not concern himself with his son's education. Instead of sending him to a school he kept him at home for an unusually long time. We find that Mikołaj stayed in the paternal home until he was nine. Finally, people prevailed upon the careless father and persuaded him to send his son to a school in the town of Skalmierz. The youngster attended the school for two years but "did not learn anything". Then the father transferred him to another school, in Lvov. In the second school, however, the boy did not fare better than in the first, because "he was already growing up". Mikołaj spent two more years there. Then the father placed the youth at Cracow University in the Jerusalem dormitory (*bursa Jerusalem*). From the *Liber matriculationis* of the University we learn that Mikołaj matriculated there in 1518, at the age of thirteen. The matriculation at such a young age should not surprise us: the University combined the functions of a high school with that of an institution of higher education. The stay at the University, again, was "of little or no avail, since he already knew the pleasures of good society".

After a one-year stay in Cracow the father took his son back home, now considering him to be a learned man. But the biographer has no illusions about that education: "as of old, he knew nothing". In his native Żórawno he idled four more years, spending most of his time hunting and fishing. In both these pastimes he was highly successful, bringing home plenty of fish, and was considered by the household to be a young man who would be able to get along well in life. But the biographer knows better. He knows that young Mikołaj is a good-for-nothing.

Then the father decided to send him to an uncle who, as he hoped, would manage to place the stripling somewhere. Most probably, he had in mind a place at the court of a neighboring lord. For this occasion he bought the son some nankeen material (*kitajka*) for a dress. But our good-for-nothing cut the material into narrow strips and then, after having caught a number of crows with a net, he made from the strips small streamers, attached them to the crows, and let the birds loose. The trick proved useful, because the crows frightened away other birds from the stackyards. But the father was hardly pleased. The young man was kept home for another year, which is what he must have aimed at, but, as a penance, he had to stay with the bailiff.

He was already twenty when the father sent him to the court of one of the dignitaries, Andrzej Tęczyński, the palatine of Sandomierz. Then, in the society of polished people he became conscious of his lack of education. The palatine used him for the writing of letters in Polish. With time, Rej managed to develop some passive knowledge of Latin, acquired the habit of reading and, being quick-witted, learned a lot from conversation. Thus, he came "*ad iudicium*" and "understood what was black and what was white". This education, however — and the author is at pains to stress it — did not go very far. There were several impediments. First of all, Rej was very fond of the pleasures of good society. Then, he liked travel and had a passion for hunting. He discovered that he had great ease at rhyming, and indulged in that activity. He was fond of music as well and knew a lot of melodies.

Thus ends the story of Rej's *Lehrjahre*. The pace of the narration is leisurely and the tone humorous. The narrator is obviously amused with his protagonist. Once we come, however, to the story of Rej's adult life we have another surprise. It was rich in events, many-sided, and a tremendously active life. Rej was a highly successful landowner who managed to build a considerable fortune. He was active in political life and served as a deputy. He set his hand to the task of spreading religious reform in Poland and, last but not least, he was a very productive writer. One cannot say that the biography does not deal with all these activities. But it deals with them in a most perfunctory way. The pace of the narration changes abruptly. It now becomes hurried, and the style resembles that of encyclopaedias or biographical dictionaries. And over and over again the author would remind us that Rej was fond of good society above all else and that he would spend endless hours carousing and merrymaking.

Rej's literary output was considerable. The biography provides us with a record of it in the form of a list of titles of his works, the more useful since he used numerous pseudonyms and some of his tracts have not been preserved. The biographer, however, wants us to believe that all this impressive amount of writing was done in spare moments only, at night, because Rej used to idle away his days in society.

We learn from the biography that he liked to be close to the court and "did not miss one diet, one public assembly or discussion of political matters" ("a żadnego sejmu, zjazdu ani żadnej koronnej sprawy nie zamieszkał" — pp. 650-651). However, except for a few instances when he served as a deputy, he would never accept a public office because he valued above all his free and "quiet" life.

The strangest of all is the biographer's treatment of Rej's business activity. Rej was born into a family of relatively well-to-do landowners. He managed to acquire several big estates, founded two towns, Rejowiec and Oksza, and did considerable money-lending. The biography does not mention money-lending but does speak about his acquisitions. It presents them, however, as if they were windfalls: he married richly, was given some estate by the kings who liked him and some by his friends. Only once is it mentioned that certain estates were bought by him as well ("i niemało był imienia tam potym przykupił" — p. 652).

In short, Rej's successes are presented here as a sort of sixteenth-century version of "how to succeed in life without actually trying". The biographer wants us to believe that the man who was primarily interested in the pleasures of good society and dallied away most of his time left behind an enormous fortune and impressively rich literary output "without actually trying".

The very long — as it was then the habit — title of the biography ends with the words which are credentials for its author: "wrote Andrzej Trzycieski, his good companion who knew all his affairs" ("który wiedział wszytki sprawy jego" — p. 654). The final words of the life sketch are: "And that was the whole story ("a toć była wszytka sprawa") of the life, deeds and affairs of this honorable Polish nobleman" (p. 655). Thus, the biography stresses that not only is its author highly competent, is an "official" biographer, but also that its text contains everything relevant for the subject. This heavy insistence makes us, as is usually the case with statements made with such a stress, rather suspicious. And makes us the more curious to find out the reason for such an unusual, highly unconventional treatment of the biographical material.

Historians of literature were quick to recognize the literary value of the biography, the vividness of its style, its immediacy and humor. The old bibliographer and historian of literature, Michał Wiszniewski, called it "a Flemish picture of domestic life of a distinguished Pole from the middle of the sixteenth century from which one cannot take one's eyes".[2] "A biography the like of which is not to be found in our old literature", wrote J. I. Kraszewski[3] a hundred years ago. "Das ausgezeichnete literarische Porträt, die köstliche Biographie

[2] Michał Wiszniewski, *Historia literatury polskiej*, VIII (Cracow, 1851), p. 142.

[3] J. I. Kraszewski, *Dziś i lat temu trzysta* (Wilno, 1863), p. 5.

Reys", the distinguished historian of literature, Juliusz Kleiner characterized it.[4] "A literary work of great importance", recently stated Tadeusz Ulewicz, a specialist in the period.[5] It is one of the most popular among sixteenth-century Polish texts, as is attested by numerous reprints. In the twenties there were lively polemics concerning its authorship, a question to which we shall return later. But, strange to say, its idiosyncratic choice of details and distribution of accents were taken for granted. The biography was treated as the outpourings of a naïve man who simply did not know better.

It should be said from the outset that our life sketch is obviously laudatory. Any double-dealing of the author with his protagonist, any attempt at presenting him as a rather funny or controversial figure is out of the question here. The biography depicts Rej as coming from honorable stock, stresses that he himself was a highly honorable gentleman, emphasizes his kindness and friendliness towards people, and presents him as a man very much liked and appreciated by everybody. And yet, at the same time, it dwells on amusing and trifling details, plays down the elements of effort, systematic work, and striving in his life, and belabors the reader with repeated statements that he was, above all, a gentleman of leisure and was spending most of his life enjoying good society. Why this insistence?

2. REJ — THE "HONEST" MAN

Stylistic analysis provides us, as it seems, with the key to the enigma. The epithet most often used in reference to Rej, his family, and his works is the one *poćciwy* (as Rej usually wrote the adjective, although he knew also the present-day form *poczciwy*). Today *poczciwy* has a slightly derogatory, condescending meaning. In present-day usage the person characterized by means of such an apposition is a decent man but a simple soul at the same time. This is, however, a modern development. As late as the middle of the nineteenth century, Fredro in his memoirs could use it, as the context proves, without having in mind any condescending note, as an adjective meaning "decent", "honorable".[6] It had that same, a decidedly positive meaning in Rej's time as well.

This apposition *poćciwy* is repeated in the text over and over again. It appears for the first time in its title: "*Żywot i sprawy poćciwego ślachcica polskiego*

[4] J. Kleiner, *Die polnische Literatur* in the series *Handbuch der Literaturwissenschaft*, ed. O. Walzel (Wildpark-Potsdam, 1927), p. 10.

[5] Tadeusz Ulewicz, "Żywot i sprawy poćciwego ślachcica polskiego, Mikołaja Reja", *Studi in onore di Ettore Lo Gatto e Giovanni Maver* (Firenze, 1962), p. 651.

[6] "*I teraz nie rzucę jego nazwiska wzgardzie poczciwych ludzi*", Aleksander Fredro, *Trzy po trzy*, ed. H. Mościcki and W. Borowy (Cracow, 1949), p. 45.

Mikołaja Reja." It reappears in the first sentence of the text proper: "Ten to Mikołaj Rej wyszedł był ze staradawnego a poćciwego domu, które zawżdy Rejmi zwano" (p. 645). It emerges twice, still with reference to Rej's family, a few sentences later: "A był to naród tych Rejów zawżdy cichy, skromny, poćciwy, nie bawiąc się nigdy żadnymi świeckimi sprawami, tylko zawżdy spokojnego a poćciwego żywota swego ślacheckiego używali" (p. 646). And it is used again when the biographer characterizes our protagonist's father, Stanisław: "człowiek pobożny, poćciwy a spokojny" (p. 647).

Two more times it is applied in the text to Mikołaj Rej. At one moment he is characterized as "człowiek poćciwy, zachowały, dworski" ("an honest, proved, polished man" — p. 655). The word appears again in the already-quoted final sentence of the text: "A toć była wszytka sprawa żywota, postępków i spraw tego poćciwego ślachcica polskiego" (p. 655). Thus, three times Rej himself is called *poćciwy*, and two times it occurs in strategically important places: in the title and at the very and of the text.

The qualification *poćciwy* or *poćciwie* is used as well in the bibliographical part of the sketch, whenever the author refers to Rej's works dealing with nobility:

Pisa też potym *Zwierzyniec* stanów ślacheckich, którzy na ten czas żywi byli, poćciwie krótkimi słowy.

Pisał księgi *Żywota człowieka poćciwego*, rozdzielony na trzy wieki jego, ... jako się ma poćciwy człowiek, na każdym stanie będąc, w swej powinności zachować.

Napisał też ... *Przemowę do poćciwego Polaka stanu rycerskiego.*

... epitafia ludziom poćciwym ... (pp. 653-654).

These quotations are telling in one respect. All four of them show clearly that in the author's mind *poćciwy* has obvious class connotations, referring to nobility. In this respect its meaning here is analogous to the traditional French *honnête homme*.

The very title of the biographical sketch starting with the words *Żywot i sprawy poćciwego ślachcica polskiego Mikołaja Reja z Nagłowic* harks back to to the largest part of the whole collection in which it was published, to the *Żywot człowieka poczciwego.* And in that large treatise as well we find copious repetitions of *poćciwy* and cognate words. Two quotations should suffice to make clear the ample use Rej made of this device:

Już wedle obietnic bożych wszytko się to na nim wypełnić musi, co Pan poćciwemu a pobożnemu człowiekowi obiecać raczył, bo bogactwa i poćciwosci w domu jego zamnożyć się muszą, bo takiego już wszyscy ludzie czczą, ważą, a poćciwość mu wszędy działają i bez słusznego bogactwa, wedle stanu swego, dom jego być nie może, gdyż wszytkiego pomiernie a poćciwie używa, a k temy, iż go wszyscy wspomagają.

With these words end the fifteenth chapter of the Second Book. And the following chapter starts thus:

Jużechmy się dosyć nasłuchali o powinności poćciwego człowieka, a zwłaszcza już tego, który wyszedszy z płochej młodości swojej a postanowiwszy staniczek swój, w którym go kolwiek Pan Bóg, powoławszy, postanowić raczył, bądź to na jakim urzędzie, bądź też w tym wolnym bez urzędu, jako żywot swój poćciwy stanowić i w nim się pobożnie zachować ma, i jako wszytki przypadki cielesne w sobie skromić ma, i jako się cnotami i sprawami rozmyślnymi a poćciwymi zdobić ma, i w nich się zachować ma, i to, co człowieku ućciwemu należy, jako się we wszem wedle powinności swej rozważnie i pobożnie zachować a sprawować ma, jakoby ni w czym nie obraził a niczym nie zaćmił jasnego a poćciwego stanu swego (pp. 353-354).

This sort of repetition of the key word and its cognates is a typical device of medieval sermons. Here is a characteristic example from a fourteenth-century popular Latin sermon written in Germany:

Nota quanta sit duritia peccatorum, quoniam tam pauci sunt, qui volunt agere penitentiam. Nota etiam maximam dei misericordiam, quia tantum modicam fecerat iste peccator poenitentiam in templo et tamen dixit dominus: descend. hic justif. Nota circa istam materiam, quod dominus octo genera penarum preparavit peccatoribus qui penitere nolunt et ista ponit Anshalmus, quod sunt haec: vincula, damnum, verbera, talio, ignominia, servitus et exilium et mors. His generibus penarum puniuntur homines excedentes leges humanas temporaliter, sic et similiter secundum leges divinas puniuntur eternaliter nisi peniteant in hac vita.[7]

The same device was frequently used by the great Italian popular preacher, St. Bernardino da Siena:

Terzo, conviene che 'l giudice sia savio per modo che non si lassi ingannare, né mai si rimuova né per lusénghe né per minaccie; però che so' di quelli che sanno tanto ben dire, che fanno rimuòvare il giudice il quale ha a giudicare; et anco so' di quelli che ogidí per le cose le quali Idio lo' manda per vero giudicio, che essi dicono: — O Dio, se tu fussi qui, io ti mostrarei con ragione che tu non fai bene a darmi questo giudicio. — A Dio sta a giudicare tutte le creature umane: non sta a te, giudice, di giudicare il religioso. Elli è posto a ordine tutti quelli che hanno a giudicare, e chi elli non hanno a giudicare.[8]

Polish medieval preaching also knew this device. Here is an example taken from the *Gniezno Sermons*:

A też ci nasz Kryst miły jest ci się on w ten to czas był narodził, gdyż ci dnia przybywa, a nocy ubywa, tegodla, iże któryć człowiek weń wierzy, ten ci będzie wezwan ku wiekuje światłości krolewstwa niebieskiego. Też ci Kryst jest ci on pozwan był

[7] Quoted after: Anton Linsenmayer, *Geschichte der Predigt in Deutschland von Karl dem Grossen bis zum Ausgange des vierzehnten Jahrhunderts* (München, 1886), p. 467.
[8] San Bernardino da Siena, *Le prediche volgari*, ed. Piero Bargellini (= *I Classici Rizzoli*, Milano-Roma, 1936), p. 271.

światłością przez to, iż ci on niebo i ziemię oświeca. A zaprawdęć by ten to człowiek był barzo grzeszny, jenże by on od te to światłości nie był oświecon w jego duszy. A przez to któryć człowiek miedzy nami chce być w jego duszy tą to światłością oświecon, prosiż ty dzisia Krysta, jenżeć się jest był jako dzisia narodził.[9]

One has to distinguish such a case from various repetitions discussed in classical rhetoric. There, the repetition served, as in *geminatio*, the purpose of heightening the emotional temperature of the text (for instance in Cicero: "excitare, excitare ipsum, si potestis, a mortuis"); or more often, the repeated word was used as a mark of the division of the speech, appearing at the beginning or the end of a certain unit of speech. That is why function words can be repeated for this purpose, as in polysyndeton.[10] The repetition with which we have to deal in these sermons is closer to classical *annominatio* but is not identical with it. The preacher wants to hammer into the heads of his hearers by the repetition of a key word and its cognates what he considers to be the basic idea of the passage. Any regularity in the distribution of the word does not come under consideration. It is a fairly primitive stylistic device, but it was revived at the beginning of the twentieth century in the massive prose of Charles Péguy.

Poćciwy being the key word of the life sketch, the main task of the sketch is to convey the idea that Rej was an "honest" man, in other words, his life conformed with the rules of behavior laid down in Rej's big treatise, *Żywot człowieka poczciwego — The Life of the Honest Man.*

The Life of the Honest Man belonged to the popular sixteenth-century genre of moralistic literature which set the standards of behavior for people in different walks of life. The best known Polish specimen of such literature is Górnicki's *The Polish Courtier* (1566), an adaptation to Polish conditions of Castiglione's famous *Il libro del cortegiano* (1528). Another example is Goślicki's Latin treatise *De optimo senatore* (1568), depicting the accomplished king's counselor, which was no less than three times translated into English (1598, 1607, 1773) and, probably, parodied by Shakespeare in the figure of Polonius.[11]

Rej's *The Life of the Honest Man* sets the standards of behavior for a Polish nobleman. A passionate champion of Calvinism in his earlier writings, this time Rej decided to address the mass of the gentry, irrespective of their denom-

[9] *Średniowieczna proza polska*, ed. Stefan Vrtel-Wierczyński, *Biblioteka Narodowa*, Series I, No. 68, 2nd ed. (Wrocław, 1959), pp. 35-36.

[10] Cf. Heinrich Lausberg, *Handbuch der literarischen Rhetorik* (München, 1960), I, §§ 608-664.

[11] Polonius was linked for the first time with Goslicki's treatise by I. Gollancz in "Bits of Timber", published in *A Book of Homage to Shakespeare* (Oxford, 1916), pp. 173-177. The point was stretched to absurd length in several papers and, more recently contested by Witold Chwalewik, *Polska w Hamlecie* (Wrocław, 1956), pp. 100-120. See also Chwalewik's comments in his recent bilingual edition of *Hamlet* (Wrocław, 1963), p. 138.

inations. In the first part of the treatise, he discussed the education of a nobleman, in the second, his adult life, and in the third, his old age. Rej had nothing in him of a systematic thinker. He was a homespun, forceful moralist, prompted by un urge to preach, and at the same time, a man obviously enjoying life and its pleasures. An amusing instance of confusion is his recipe for preparing sauerkraut, inserted into the treatise (p. 366). He was also hopelessly foggy whenever discussing general ideas. In spite of all these shortcomings, the whole treatise, although at long stretches repetitious and dull, makes vivid reading stuff, being studded with crude but vigorous satirical thumb-nail sketches and breathing hearty love for country life and its pleasures. It gives the reader a vivid idea of the ethos and mentality of "merry old Poland", especially in reflecting the self-confidence of the Polish gentry, the class which had just gained political ascendancy and was enjoying a period of considerable affluence.

The gentry was primarily the class of landowners and grain exporters. However, they based their claims for supremacy on the idea that they were the knightly class, that their basic function was to defend the country. They justified the institution of serfdom by their knightly mission: the peasant worked for his landlord, and the landlord, in turn, defended him whenever necessary.

In fact, at the time Rej was writing his treatise Poland had been enjoying relative peace. The gentry was above all the class of agricultural producers and exporters. The good prospects of the export of grain via Gdańsk (Danzig) made the cultivation of land highly profitable, and served as incentive for intensive production by means of extending the obligations of the *corvée*. At the same time, the gentry was contemptuous of any other economical activity, the activity that was typical of the burghers. A member of the gentry who would resort to trade or craftsmanship was considered to be dishonored and liable to lose the privileges of his class.[12]

The Life of the Honest Man reflects that ideology. It does not mean that Rej rigorously excludes the incentive of money from his presentation of "honest" life. He mentions money several times, and when speaking of it he makes use of one of the endearing diminutives, so typical of his style: *pieniążki*. Even when speaking of such a pastime as hunting he does not forget the profit incentive: a wolf's or fox's hide can be profitably sold (p. 368). But money is never for him the main reason for squire's work. The "honest" man should work primarily because the work on a farm is so pleasant, so enjoyable. Reviewing different

[12] Cf. Jan Rutkowski, *Zarys gospodarczych dziejów Polski* (Poznań, 1923), Part II, Chapter III; Franciszek Bujak, "Zasady polityki gospodarczej Polski w w. XVI i ich geneza", and Stanisław Kutrzeba, "Życie społeczne", in *Kultura staropolska* (Cracow, 1932).

kinds of activity, over and over again he mentions in one breath profit together with "delight" and "fun", — *rozkosz* and *krotochwila* in his vocabulary (pp. 358, 360, 361, 362, 366). Working on a landed estate provides, above all, a pleasant discharge of a surplus of energy. In conformity with these ideas, when reviewing different sorts of squire's activity, he puts gardening in the forefront. This is the sort of occupation worthy of a landlord. At the same time, the cultivation of grains, the basis of Polish agriculture, is given only a very cursory treatment. How much more suitable for the squire is the grafting of trees than the supervising of manuring, which can be done by any bailiff (p. 358).

Rej wants his squire to be active. But the kind of work he preaches is light and easy, achieved without great effort, more an entertainment than a toil. In a telling passage he condemns dangerous and difficult kinds of work characteristic of lower estates (*mniejszych stanów*), such as the work of a miner, a mariner, a man collecting wild bees' honey, or a professional soldier. To be exposed to danger or to toil in order to earn money is, he argues, vile, is a proof of greed. It is as despicable as the occupations of a dogcatcher and a hangman, or as the actions of a man who would poison for money or of a woman who would sell her daughter into "obscene slavery" (p. 228).

Rej preaches satisfaction with what one has. There is nothing reprehensible in having an estate. Wealth is God's gift (p. 237). But an urge to enrich oneself is reprehensible. One can detect in Rej's presentation of advantages gained by good husbandry the attitude of a man who is not free from greed for *pieniążki*, "dear money". But in spite of the fact that such an attitude is quite palpable occasionally, Rej the moralist is never tired of repeating that the "honest" man has to be content with his estate, no matter how modest, and that a striving after enrichment is blameworthy. The basic rule of the art of good living is to live quietly in the confines of one's estate. In order to stress the point, Rej resorts to a drastic example. Even if a squire has on his estate a peasant richer than himself he should not envy the peasant but should learn to be reconciled with such a state of affairs (p. 370). Otherwise, the squire is guilty of sinful pride. Rej preaches humility and moderation, an attitude that was in keeping with the static, quietistic model of the gentry's social ideal.[13]

Social ethics presented in *The Life of the Honest Man* basically conform with the teaching of the Catholic Church, such as were codified by Aquinas. The Church envisaged social classes as static entities. Man should be satisfied with the place on the social ladder allotted to him. He should not try to climb higher. The possession of wealth is a morally indifferent factor, and the cultivation of

[13] Well brought out in relief in Andrzej Zajączkowski's "Cadres structurels de la noblesse", *Les Annales*, XVIII (1963), 94-95.

land is the noblest kind of work.[14] Rej's social ethics differ from those of Aquinas in one important aspect. Aquinas considered the contemplative life to be the highest possible type of human activity. Rej's savage mockery of contemplative orders, in which he indulged in his earlier works, implies a total rejection of these high claims. Rej tried to avoid giving his *Life of the Honest Man* a confessional stamp, and there is no trace of that mockery in his treatise. But it does not mean that he revised his stand. There is no place for *vita contemplativa* in Rej's picture of the life of his "honest" squire.

In 1904 Max Weber published his famous work *Die protestantische Ethik und der Geist des Kapitalismus*, in which he propounded the thesis that Calvinist ethics were a mighty stimulus for modern capitalistic activity. The thesis has been vividly debated even since.[15] It is obvious that Rej's treatise brings nothing to support Weber's thesis. But if we turn to Rej's actual life the picture is quite different.

3. REJ — THE BUSINESSMAN

It is a shame that we have no good biography of Rej. Several books on Rej, among which Aleksander Brückner's massive volume *Mikołaj Rej, Studium krytyczne* (1905) occupies justly the place of pride, deal exclusively with his works, and for biographical data rely mostly on our life sketch, published under the name of Trzecieski. Somewhat more biographically oriented is the latest among books on Rej, that by Stanisław Wasylewski.[16] But it does not lead us any further. Conceived as a short, popular book, it is based on scanty research and displays only a superficial knowledge of both the period and the sitter.

With most sixteenth-century Polish writers the difficulty in writing their biographies consists in the paucity of available source material. With Rej it is otherwise, and the side of his life which is most richly documented is his economic activity. We have at our disposal a major collection of documents, *Materiały do biografii Mikołaja Reja*, published by Zbigniew Kniaziołucki.[17] It contains a rich crop, more than eight hundred documents, mostly from Cracow and Lvov archives, either published in full or summarized. And they deal primarily with

[14] Adriano Tilgher, *Work*, transl. D. C. Fisher (London, 1931), pp. 31-42.

[15] A recent lively review of the debate can be found in the article by Herbert Luethy "Once Again: Calvinism and Capitalism", *Encounter*, January 1964, pp. 26-38.

[16] *Rej z Nagłowic*, in the Series *Wielcy ludzie*, Vol. II (Lvov, 1934).

[17] *Archiwum do Dziejów Literatury i Oświaty w Polsce*, VII (Cracow, 1892), 241-641. A few additional documents were published later by Teodor Wierzbowski in *Z wieku Mikołaja Reja* (Warsaw, 1905), Part II, pp. 6-13.

different aspects of Rej's business activities. According to Brückner, some more material could have been found as well in the Przemyśl archives.[18]

The documents published by Kniaziołucki throw a lot of interesting light on Rej. Their large number is revealing. They testify to business activity conducted on an impressively large scale by a man seething with energy and determined to build up a large fortune. Rej is conducting vast credit operations, lending and borrowing large sums of money, buying villages and founding new ones, exchanging estates, founding on his territories two new towns, Oksza and Rejowiec, and buying a house in Cracow. Some of his operations were conducted in grand style. Thus, to quote one instance, on April 13, 1552, he lends to Jan Tęczyński, the palatine of Sandomierz, the sum of 1.500 florins, and takes as security Tęczyński's estate, Blinów. No lesser a figure than "*Sigismundus Augustus*, D. *g. rex Poloniae*"[19] functions as notary public in the transaction. This seething activity involves him in numerous lawsuits and forces him to spend most of his life in travels.

These activities were eminently successful. Rej built up a large fortune. His vast estates were concentrated in three different regions: in the south-eastern part of the Commonwealth, the so-called Ruś Czerwona, in the Chełm region, and in Western Małopolska. Two of his sons married heiresses of aristocratic families: Mikołaj took as wife a daughter of the Wilno palatine, Jan Hlebowicz, and Andrzej married a daughter of the chancellor Walenty Dembiński.[20]

The documents collected by Kniaziołucki allow us to check step by step the data on Rej's acquisitions of different estates given in the biographical sketch. This sketch presented as the first step in Rej's building up of his fortune his marriage to "a certain Miss Kos from Sędziszów", a wealthy heiress owing to the provident care of her uncle, the primate Andrzej Boryszewski, "who bought a lot of estates for his relatives". The wife brought him, we learn there, two estates in the Chełm region, Kobyle and Siennica. This is borne out by a document in Kniaziołucki's collection (No. 283), dated December 20, 1531, in which Rej bequeathed Topola to his wife as a life estate, and she, in turn, bequeathed to him "bona in terra et districtu Chelmensi".

As the second source of Rej's wealth the biographical sketch indicated the favors which the court had showered on him. Both Sigismund the First and his wife Bona, as well as their son Sigismund Augustus, were friendly toward him. The old king granted Rej a yearly pension and the privilege to collect the annual tax from the town of Skalmierz. He gave Rej an estate as well. From a docu-

[18] *Mikołaj Rej*, p. 15.
[19] Kniaziołucki, No. 554.
[20] See the genealogical tree in Kniaziołucki, p. 613.

ment[21] we learn that it was the village of Temerowce. The document states that Rej was given the estate for some unspecified merits ("pro meritis in nos"), and was registered under the title: "Donatur villa Temerowcze g. Nicolao Rey vati Polono alias Rymarzowi". This makes probable the assumption that this royal gift was the first known Polish literary prize.

From Sigismund Augustus, the life sketch informs us, Rej obtained another pension, based on Chełm toll, as well as an estate, Dziewięciele. About the latter favor the text is quite specific. The village, we learn there, belonged to Mikołaj Odnowski, but after his death had to become the property of the royal fiscus *iure donatorio*. Odnowski, being friendly toward Rej, allowed him to take the estate into his possession during his lifetime.

The documents show this transaction in a rather different light. It turns out that Rej had business relations with Odnowski for a long time, at least from 1536 (No. 319). In 1544, Odnowski instigated a lawsuit against Rej, who had not paid him for the grain he had bought from Odnowski's three estates. Rej did not appear in court, and Odnowski obtained a verdict against him (Nos. 387-389). Later on, this relationship must have been straightened out, because in 1546 Rej borrowed from Odnowski 300 florins, pawning for them his estate Kołodziejów (No. 408).

Dziewięciele, or Dziewięcioły, was not a property of Odnowski but belonged to the royal domain. Together with parts of two other estates it was leased to Odnowski for life tenure. By a document, dated March 24, 1552 ,Sigismund Augustus donated the estate to Rej, but only after the death of its tenure holder, Odnowski (No. 553). Rej, however, paid off the tenure holder. He took possession of the estate at once, and promised to pay 212 florins to Odnowski every year (No. 574). Thus, what the biographical sketch describes as a free gift of a friend was, in fact, a typical business transaction.

The sketch wants us to see another such friend of Rej, an open-handed donor in the person of Piotr Bystram. Bystram was childless. In 1541, he bequeathed his estate to his father-in-law, Myszkowski (No. 370). Probably, as was surmised, he wanted to exclude from the inheritance some relatives whom he did not like.[22] Half a year later, he annulled that bequest (No. 379). On January 3, 1548, he made a new one, this time in favor of Rej (No. 448). The day before, however, Rej and Bystram had signed an agreement by which they made void all their previous transactions under the pledge of a vast sum of 10.000 florins, "sub vadio decem millium flor. pec." (No. 447). We can safely surmise from all this that Rej must have lent Bystram a large sum of money, and that Bystram

[21] Kniaziołucki, No. 410; dated February 13, 1546.
[22] Bronisław Chlebowski, "Mikołaj Rej w świetle nowych świadectw", *Pisma* (Warsaw, 1912), II, 54.

made Rej his inheritor in return for the annulment of his debt. Again, what is presented in the biographical sketch as a liberal gift turns out to be a business transaction.

The sketch, after having mentioned the founding of the town of Oksza, adds that later on Rej "bought a lot of land in the vicinity of the town" (p. 652). Otherwise, it wants to present Rej as having become rich primarily by means of free gifts. It states that the Archbishop Gamrat leased to Rej two towns, Kurzelów and Biskupice (a transaction which was not attested in Kniaziołucki's collection), and, when writing on royal favors, adds that "all the lords (*wszyscy panowie*)" favored Rej and "gave him a lot (*i wiele mu dawali*)" (p. 651). These "free gifts", we can be sure, were of similar character to those from Odnowski and Bystram.

In 1893, shortly after Kniaziołucki's collection had been published, Bronisław Chlebowski confronted its data with the biographical sketch in a paper, later reprinted in the collected edition of his works. Chlebowski, as befitted the editor of the great *Słownik Geograficzny*, had a good orientation in the social and economic conditions of sixteenth-century Poland. At the same time he had a penchant for far-reaching psychological explanations which sometimes assumed the character of irresponsible fantasies. Moreover, his paper contains a serious factual error. Not able to disentagle the biographical data of the two Andrzej Trzecieskis, father and son, he assumed that the biographical sketch was written by the older Trzecieski.[23]

Chlebowski noted wide discrepancies between the biographical sketch and the data provided by the newly published documents. Attributing the biography to the old Trzecieski who, had he lived so long, must have been almost seventy by 1568, he explained the discrepancies as the effect of the failing memory of the old author of the sketch who neither managed to put order into his rambling reminiscences, nor cared for checking his facts. The sketch, according to him, was an interesting but primitive and senile work.[24] Thus, a wild, arbitrary "psychology" took care of the whole problem.

The relevance of the documents was discussed again in 1905 by the greatest authority on Rej, Aleksander Brückner, in his *Mikołaj Rej*. Brückner was a scholar who combined tremendous erudition with unbridled emotionalism. He was especially fond of homespun old Polish works which would bring the scholar in the closest possible touch with the mentality and habits of the period and would allow him to revel in the riches of the old language. Rej, together

[23] One should add that only some twenty years later was the exact date of the death of the senior Trzecieski, 1547, established; see Jerzy Krókowski, *Andrzej Trzecieski, poeta-humanista i działacz reformacyjny* (Warsaw, 1954), p. 15.

[24] *Pisma, ut supra*, pp. 38, 48.

with the seventeenth-century poet, Wacław Potocki, fulfilled both conditions as no other Polish writers did. Thus, Brückner was enamored of Rej.

When discussing the documents he went to special lengths to stress that none of them revealed facts which would compromise Rej's integrity, and he quoted with approval the editor Kniaziołucki's words to this effect.[25] He noticed that the documentary version of the transaction with Bystram was incompatible with that of the biographical sketch and he explained that incompatibility away as a lapse of Trzecieski's memory. The documents reveal several lawsuits, mostly with the clergy. Brückner dismisses them as "usual gentry conflicts". The documents, Brückner concludes, give the picture of Rej's business activities, but don't throw any light on Rej the man, and, especially, on the writer.[26] Returning to the subject a few pages later, Brückner asserts that we cannot go in our knowledge of Rej the man beyond that which we find in the biographical sketch: "We cannot either enlarge or deepen Trzecieski's story filling these frames."[27] Like the author of the sketch, Brückner was bent on proving that his beloved author was an "honest" man, and, for this reason, he was so prone to swear by the text of the sketch, dismissing the testimony of the documents as irrelevant for a literary historian.

The authority of the great scholar prevailed the more easily, since the documents written in a barbaric Latin and dealing with the intricate details of sixteenth-century economic life are not attractive reading matter.

4. THE SKETCH AS A TESTIMONY OF HONESTY

In fact, the biographical sketch, far from being written with easy abandon, is a carefully calculated piece of work giving a highly stylized picture of Rej's life. And it is not difficult to find the reasons for such a stylization. Rej's *Life of the Honest Man* preached a certain way of life marked by moderation and satisfaction with one's own station in life. Rej's "honest" nobleman was a man of leisure above all. At the same time, Rej's actual course of life was so different from the ideal preached by Rej, the moralist. With tremendous zest Rej plunged into business activity, was motivated throughout his life by a passion for enrichment.

Thus, by publishing his treatise he exposed himself to objections that he was preaching the way of life opposed to that which he himself had been following. He exposed himself the more easily, since he was by no means an obscure writer. He was a public figure, a leading champion of the Reform movement,

[25] *Mikołaj Rej*, p. 15.
[26] *Ibid.*, pp. 15-16.
[27] *Ibid.*, p. 19.

a noted writer, occasionally, mostly in the years 1556-1564, a deputy actively participating in the parliamentary life of the country,[28] and a frequent visitor to the court. His renown as a writer is attested to by such authors as Kochanowski, Bielski, Paprocki, and Klonowic, to mention only the most representative among sixteenth-century writers. As an eminently social, agile, and prosperous man he must have been known to wide circles of his contemporaries.

As a matter of fact, Rej's private life was being attacked in print. We may easily dismiss the testimony by a Catholic polemist, Father Józef Wereszczyński, who in a pamphlet published in 1585 under the succulent title *Gościniec pewny niepomiernym moczygębom a obmierzłym wydmikuflom świata tego* (*A Sure Gift for Intemperate Bibbers and Loathsome Tosspots of This World*) presented Rej as a monster of gluttony and drunkenness. Writing some fifteen years after Rej's death, in the atmosphere of victorious Counterreformation, Wereszczyński, a not too scrupulous pamphleteer, was eager to obliterate the memory of the once-famous Protestant champion whom he plagiarized elsewhere. Moreover, the issue raised by him is without great relevance. Most probably, Rej's eating and drinking habits were not too refined but, after all, an indulgence in the pleasures of the palate was not incompatible with the idea of the quietist, leisurely life Rej was preaching. Rather the opposite was true.

Far more relevant is the testimony unearthed at the beginning of the nineteenth century by a learned antiquarian, Hieronim Juszyński. In his *Dykcjonarz poetów polskich* he quotes from a pamphlet by a Protestant writer, Jan z Woźnik, *De apostolica Jesu Christi doctrina in Regno Poloniae predicata*, published in the very year 1568 in which Rej's *Life of the Honest Man* was printed, and obviously referring to it. This is what he had to say, as translated from Juszyński's Polish translation:

> I cannot praise Rej, because he did not help but impeded the spreading of the holy truth. Our enemies looking at his doings pointed at him and used to say: you would recognize them by their fruits. Violent in his affairs, always unpremeditated in their fulfillment, he wrote the life of a Christian man, and his doings were not human.[29]

Later on, in the same passage Jan z Woźnik mentions as two instances of Rej's misdeeds his action against Cracow Franciscan monks and another against a certain Dunin from Karwice. We know nothing about these litigations from other sources.

At the end of his discussion of Rej, Juszyński returns to the same pamphlet and quotes Jan z Woźnik again, this time in the original Latin. Addressing Rej and reproaching him with his ignorance, our pamphleteer then goes on: "Pul-

[28] Stanisław Bodniak, "Mikołaj Rej na sejmach", *Pamiętnik Literacki*, XXV (1928), 70-81.
[29] *Dykcjonarz poetów polskich* (Cracow, 1820), II, 116.

chrum depinxisti hominem, pictor foedus: aliosque ad perfectionis littus dirigis, qui adhuc in delictorum fluctibus versaris."[30]

Juszyński's work is our only source of information about the pamphlet, since no extant copy of it is known today and no scholar after Juszyński had it in his hands. An allusion to a detractor of Rej, probably to Jan z Woźnik, is to be found in a short poem in the second, posthumous edition of Rej's collection of didactic poems, his *Zwierzyniec* (*Bestiary*), published in 1574. The writer of that poem mentions a certain "sophist" who in his book "shouts mad things, mentioning some of them, already slanders the deceased".[31]

"Would some malevolent people refuse Rej the right to moralize because of his way of life?" — asks Brückner in connection with the above-quoted words.[32] The answer seems to be: yes. The discrepancy between Rej's moral teaching exposed in his last book and that which has been known about his life was so obvious that it invited criticism. The biography was written in order to forestall it, to bridge the gap between Rej the moralist and Rej in flesh and blood.

Rej preached the morals of containment, while busy with building up a great fortune. The biographical sketch wants us to believe that that fortune was not the effect of lifelong efforts but primarily the result of happy occurrences. It presents Rej as a general favorite showered by all and sundry with gifts of estates. The method was naïve, but, after all, the case was quite difficult.

Rej preached a quietistic, leisurely way of life, while his own life was full of bustle and effort. Here the biographer had two lines of defense. On the one hand, when dealing with Rej's adult years he was indefatigable in repeating that Rej was so inordinately fond of good company, that he spent long hours carousing. On the other, he dwelt upon the only period in which Rej was fully a man of leisure, a stripling who idled his life away, one who did not want to learn, and spent his days hunting, fishing and performing practical jokes. It was the only period of his life in which he lived according to the rules he himself set in his old age as befitting an "honest" man, his only "honest" period.

In short, the biographical sketch was conceived as a testimony of "honesty" for Rej, "honesty" in the sense defined in his big treatise. If we look at it from this point of view, all its seeming oddities and vagaries turn out to be calculated strategic moves.

The biographer had an axe to grind. He was bent on proving that there was no discrepancy between Rej's actual life and the teaching of his late treatise. The

[30] *Ibid.*, p. 131.

[31] "Obłędliwości wrzeszczy, niektóre mianując, już zmarłego sromości." I am quoting Brückner, "Wydawnictwa Rejowe", *Reformacja w Polsce*, I (1921), 253.

[32] *Ibid.*, p. 253.

treatment of the confessional aspect of Rej's life in the biography confirms this view. As we have seen, Rej's treatise was addressed to the gentry at large and not only to Protestants. Because of this Rej played down denominational questions there. We find the same attitude in the biographical sketch. Rej's belonging to the Protestant camp is not negated there. But it is not emphasized either. It is mentioned in connection with Rej's literary activity which started "when the holy truth of Lord's Gospel had come to Poland" (p. 652). But the man who became famous above all as the passionate champion of the Protestant cause in Poland is never paid a tribute there. We know from other sources that Rej founded on his estates at least four Protestant churches.[33] Certainly, that was something to boast about. But there is not a word about it in our sketch. Not a word either on the part played by Rej in Protestant synods.[34]

Thus, the life sketch fits the big treatise as the glove fits the hand. Instead of being a rambling, naïve prattle it is, in fact, a slyly calculated work. But how artfully it is done! With what ease, verbal verve, and naturalness it assumes the air of just a simple, disorderly narration. Certainly, it is a biographical document of very dubious value. But on another level we must judge it as a work of verbal art.

5. THE PROBLEM OF THE AUTHORSHIP

The main object of this paper is to analyze the basic tendency of the biographical sketch and to show how it determined the seemingly fanciful, in fact very careful and ingenious choice of details. It seems, in addition, that its findings can bring us nearer the solution of the problem of its authorship.

The problem was raised forty years ago by a linguist, Henryk Gaertner, who in a careful, detailed, and well documented analysis proved that the language of the biographical sketch is identical, as concerns vocabulary, phraseology, and typical syntactic structures, with the language of works coming from the pen of Rej. And he drew the conclusion that the sketch was Rej's autobiography.[35]

Gaertner's argument is strong as long as he discusses strictly linguistic matters. When he, however, tries to find out the psychological reasons for the mystification he becomes disarmingly naïve. Rej, he asserts, wanted to conceal his authorship because of the elements of self-praise which, although moderate,

[33] H... M..., "Rejowie z Nagłowic jako członkowie polskiego Kościoła ewangelickiego", *Z wieku Mikołaja Reja* (Warsaw, 1905), Part I, pp. 50-51.

[34] Stanisław Zachorowski, "Najstarsze synody arian polskich", *Reformacja w Polsce*, I (1921), pp. 216-217.

[35] Henryk Gaertner, *Ze studiów nad językiem polskim w XVI wieku* (*Kto jest autorem życiorysu Reja*?) (Warsaw, 1925).

could have been embarrassing for him.[36] A few pages later, commenting on the heavy stress in the sketch on Rej's dislike of work he ventures another psychological motivation: "Rej probably wanted to leave people an authentic testimony", and "by a candid confession of certain peccadillos he wanted to counter the unfavorable rumors about his person which were being spread by Catholic priests".[37]

Gaertner's thesis was violently attacked by Brückner.[38] In order to understand the tenor of Brückner's polemics one has to bear in mind that he was a very temperamental man, that he considered the works of Rej as his domain, and that he strongly resented a hypothesis which would cast doubt on the integrity of his beloved author. Brückner did not deny that the biographical sketch was written in a language strongly resembling that of Rej but, according to him, that did not constitute a proof that Rej was its author. Trzecieski, he assumes, when paying a tribute to Rej decided to present it in Rej's style, and wrote a pastiche.

The polemics went on, with both sides persisting in their views but without adducing any new arguments.[39]

Later Stanisław Hrabec added one more small but telling linguistic argument for Rej's authorship. In the sketch we have a Ukrainian word for hemp, *płoskonki*, which was used in other works by Rej. Rej came from the eastern part of the old Commonwealth, and words of Ukrainian origin occur quite often in his vocabulary;[40] this is not the case with Trzecieski, who came from a more western region.

Recently, Trzecieski's meagre Polish literary output was for the first time collected together in a good critical edition. It was provided with a detailed paper by Stanisław Bąk, discussing Trzecieski's Polish. Bąk found some more linguistic arguments supporting the thesis of Rej's authorship of the sketch. One of them is especially strong: in Rej's works we find a profusion of pluperfects. We find many of them in our biographical sketch as well. However, they are rare in works of Trzecieski's certain authorship.[41] Accordingly, the sketch is reprinted there among works about which uncertainty exists concerning Trzecieski's authorship.

[36] *Ibid.*, p. 68.
[37] *Ibid.*, p. 70.
[38] "Spory o autorów", *Przegląd Współczesny*, XV (1925), 144-148. See also Brückner in *Przegląd Warszawski*, 1925, III, 225-226, and *Reformacja w Polsce*, IV (1926), 229-230.
[39] Gaertner, "O językowe sprawdziany literackie", *Pamietnik Literacki*, XXV (1928), 347-380, and the rejoinders of both polemists, *ibid.*, pp. 673-686.
[40] Stanisław Hrabec, *Elementy kresowe w języku niektórych pisarzy polskich XVI i XVII w.* (Toruń, 1949), pp. 26-27.
[41] Stanisław Bąk, "Język pism Andrzeja Trzecieskiego", in A. Trzecieski, *Pisma polskie*, Part I (Wrocław, 1961), p. 119.

The edition is provided with good indexes of words: one for works of Trzecieski's certain authorship, and another for those of dubious authorship. These indexes make relatively easy the task of comparing the vocabulary of the sketch with that of the works of Trzecieski's undoubted authorship. It turns out that while in the sketch we have the form *żadny* (none), Trzecieski writes *żaden*. Three times in the sketch we meet the adverb *snadnie* (quickly), nonattested in the writing of Trzecieski, who uses in that function *prędko* or *prętko*. In the sketch we find eight instances of *zawżdy* (always), in Trzecieski's writings, twice *zawżdy* and fourteen times a more recent form of the adverb, *zawsze*. Rej was born in 1505, Trzecieski, about a quarter of a century later, and it is only natural that this difference in age is reflected in their respective vocabularies and grammatical models.

Gaertner's findings failed, however, to convince the majority of Polish historians of literature. They were crisply rejected by, among others, the great authority on the period, Julian Krzyżanowski;[42] and, quite recently, Trzecieski's authorship found a spirited defender in the person of Tadeusz Ulewicz. His paper,[43] coming from the pen of an expert in sixteenth-century Polish literature, vigorously written, marshalling all the previously used arguments, and adding some new ones, is the best sustained effort at defending the traditional attribution, and constitutes a good point of departure for the discussion of the problem of the authorship.

Following Brückner, Ulewicz contends that Trzecieski decided to write Rej's biography in Rej's style, "*po nagłowsku*". The idea of such a pastiche, he argues, was not incompatible with the literary circumambience of sixteenth-century Poland. As proof, he mentions Kochanowski, who reconstructed the lost parts of Cicero's translation from the Greek of Aratus' *Phaenomena*, "trying — as he asserts — to adapt his reconstruction to the style and tone of the language of the translator", and supposedly succeeding in his attempt.[44]

These assertions, however, must be due to a misunderstanding: they are all at variance with the facts. Kochanowski did not try to recreate "the style and tone" of Cicero's translation. He provided his own Latin text for the missing parts of Cicero's translation, about one third of the whole. As for the rest of the text, he incorporated preserved fragments of Cicero's translation, but after having revised some and after having substituted his own version for some others which he considered to be inferior. And he made no bones about it. In the preface to his version he stated flatly that he considered Cicero's translation

[42] See his introductory essay to his above-mentioned edition of Rej's *Żywot człowieka poczciwego*, pp. LXVIII-LXIX.
[43] "Żywot i sprawy poćciwego šlachcica polskiego Mikołaja Reja", *Studie in onore di Ettore Lo Gatto e Giovanni Maver* (Firenze, 1962), pp. 651-662.
[44] *Ibid.*, p. 661.

inadequate ("Quo in genere noster Cicero, etsi is quoque, mea quidem sententia, Aratum non adaequat").[45] Because of this he rejected several Ciceronian passages and replaced them with his own renderings.[46] No wonder that he was upbraided for arbitrariness by a modern classical philologist.[47]

Kochanowski did not respect Cicero's version in spite of the fact that Cicero was revered by humanists as the master of Latin style. Trzecieski was an educated man who could write in smooth, correct Latin and Polish. The pastiche hypothesis wants us to believe that while writing Polish prose he resorted on purpose to the vigorous but unpolished, undisciplined prose of Rej, and did it not in order to make fun of Rej, to deride him — we have in old Polish literature a few very rudimentary attempts at ridiculing the language of uneducated people — but in order to pay him homage. And we would have to believe that he did it in an unusually meticulous way, imitating even Rej's minor idiosyncrasies, so that a careful, detailed analysis could not disclose one single discrepancy between the language of the biographical sketch and that of Rej. For a humanist reared on classical writers Rej's prose must have sounded a bit rough, unpolished, primitive. The appreciation of the specific artistic qualities of a primitive style came with the Preromantic historicism. It would be anachronistic to assume its presence in sixteenth-century Poland. If one wants proof, Kochanowski would provide it. He translated into Polish verse fragments from the *Iliad* and from Euripides' *Alcestis*. It is a far cry from Homer's Greek to that of Euripides. But in vain would we look in Kochanowski's translation from the *Iliad* for any attempt at the archaization of the language. He translated both texts into the same Polish, his own Polish.

As it has been revealed lately, the Frenchman Peter Statorius makes especially frequent use of quotations from Rej in his *Polonicae grammatices institutio* (1568). Ulewicz concludes from this fact that Rej's language "must have been considered in the middle of the sixteenth century to be the model of literary Polish".[48] This is a far-fetched conclusion. Certainly, Rej was widely read — his *Postylla* ran into five editions — and as such must have been influential. It is an obvious fact, however, that the Polish of educated writers among his contemporaries, such as Górnicki, Orzechowski, Kromer, differed widely from Rej's Polish. If Rej's text was quoted in a grammar, it means only that he was considered by one grammarian to be typical of the current usage of the language, and nothing more.

[45] Joannis Cochanovii, *Carmina latina*, ed. Josephus Przyborowski (Warsaw, 1884), pp. 377-378.
[46] *Ibid.*, p. 380.
[47] Kazimierz Morawski, *Andrzej Patrycy Nidecki, jego życie i dzieła* (Cracow, 1892), pp. 377-381.
[48] *Op. cit.*, p. 660.

Ulewicz argues further that the mystification concerning the authorship would "neither serve any purpose, nor be useful".[49] The preceding analysis of the specific purpose of the biographic sketch takes care of that argument, as it seems. If we view the biography as basically a testimony to Rej's "honesty", then the mystification makes sense. In order to be of value the testimony should have been signed by somebody other than Rej, and by someone with a well-known name. Trzecieski was a known champion of the Reform movement, and his name must have carried certain weight. He might have agreed to sign the sketch out of friendship. He might have been induced by other reasons as well. As we know now, in the years between 1565 and 1574, after he had lost his patron, Mikołaj Radziwiłł Czarny, he was in financial straits. Several of his poems from these years are petitions for help addressed to different rich and influential people. His situation improved only in 1574, when he was appointed royal secretary. Four years later, he was granted a pension, sixty florins yearly.[50]

From these lean years dates his Latin epigram in honor of Rej,[51] in which Rej was called the Polish Dante, Petrarch, Homer, and Virgil. It is probable as well, although not certain, that from his pen came also a Polish epigram, published in the same collection, *The Mirror*, in which our biographical sketch appeared. There Rej was named a new Cicero and Virgil.[52] Maybe it was friendship that dictated these brazen praises.

What can we know today? In any case, these eulogies with their exorbitant and conventional praise look like a typical product of a client's muse. Not in such term is Rej the writer praised in the biographical sketch signed with Trzecieski's name.

It is only recently that we can have a clear picture of Trzecieski the writer. We owe it to a short but thorough study by Jerzy Krókowski which brought the first detailed analysis to Trzecieski's poetry, then to the edition of Trzecieski's Latin and Polish works (1958-1961), which for the first time gathered together Trzecieski's scattered and otherwise not easily available literary production. Krókowski in his analysis on the one hand stresses Trzecieski's mastery of Latin and of classical and humanist current *topoi*. On the other, he points to the lack of any individual stamp in this mostly panegyrical poetry, to its pervading banality.[53] This analysis is borne out by the perusal of Trzecieski's works. The overall impression is that of a boring commonplace versifier, and and it is enhanced by the author's ease in expressing himself. That such a man could have understanding of the highly individual, rough, and going-counter-

[49] *Ibid.*, p. 658.
[50] Krókowski, *op. cit.*, pp. 82, 103, 111.
[51] *Epigr.*, I. 29; *Carmina*, ed. Jerzy Krókowski (Wrocław, 1958), pp. 82-86.
[52] *Pisma Polskie*, Part I, p. 219
[53] *Op. cit.*, pp. 130-136.

to-humanistic precepts which is Rej's way of expression, moreover, that he would be able to counterfeit in minute details Rej's idiosyncratic style, does not look plausible.

One should add that short autobiographies are quite common in sixteenth-century Polish literature, as is proved by such works as *Vita Ioannis Dantisci*, Ianicius' *De se ipso ad posteritatem*, Gregorius Samboritanus' Latin autobiographical elegy, or the Polish autobiography with which Strykowski preceded his chronicle.

In view of all these considerations, Trzecieski's authorship of the biographical sketch seems highly improbable. It could come from his pen under one condition only: if we assume that Trzecieski was writing it under Rej's dictation.

6. CALVINISM AND CAPITALISM: REJ AS A TEST CASE

The question of authorship, interesting as it is, should not conceal from our view a much wider and historically much more important problem which the biographical sketch poses. At the end of his life, when he was old (*The Life of the Honest Man* appeared at the time when its author was sixty-four) and satiated, he embraced the official, quietist ideology of his class. But during the rest of his adult life he was indefatigable in appeasing his considerable acquisitive appetites. At the same time, he was an unremitting lay preacher, full of religious zeal and a passion for moralizing. How did he reconcile these two activities? Do we have to resort to a sort of Dr. Jekyll and Mr. Hyde type of psychology? Or, perhaps, there was nothing to be reconciled? Perhaps there was no conflict in Rej's mind?

Rej's biography, which unravels the intricacies of his financial dealings, could be an excellent test case for Max Weber's famous thesis, because it could test the theory against a far different social background. Weber wanted to prove his thesis through an analysis of the urban economy of Calvinist Geneva and of English Puritans. In the case of Rej we could study the interrelation of Calvinism and the acquisitive passion against the background of the rural economy of sixteenth-century Poland. And Rej seems to be an ideal subject for such a study. On the one hand, we have rich documentary material concerning his economic activities. On the other, he was a prolific writer who in his voluminous literary production expressed his views on the world, on Christianity as he understood it, and on man's duties. What is needed is a scholarly analysis that would bring the two together and thus give us a deeper insight into Rej's mentality. In such an analysis the biographical sketch, one surmises, would play a not unimportant part, but only as an ingenious piece of special pleading.

SIENKIEWICZ VERSUS KRASZEWSKI: OBSERVATIONS ON NOVEL-WRITING

DAVID WELSH

University of Michigan

As a great novelist, Henryk Sienkiewicz (1846-1916) owed a substantial debt to his literary traditions. Like all writers possessed of what T. S. Eliot has called "the historical sense", Sienkiewicz wrote "with a feeling that the whole of the literature of Europe from Homer and within it, the whole of the literature of his own country has a simultaneous existence and composes a simultaneous order."[1] The exact nature of this feeling and this debt in Sienkiewicz's work is still to be studied, but during his own lifetime, Polish literary traditions were in many ways personified in his eminent predecessor J. I. Kraszewski (1812-1887). This was the writer who "delineated the path of development of the novel in Poland".[2] There can be no doubt of Sienkiewicz's respect and admiration for Kraszewski's contributions to such fields as literature, criticism, European politics and the preservation of Polish historical and cultural traditions at a time when they were threatened by the partitioning powers. In their historical novels, both Kraszewski and Sienkiewicz sought to illuminate the grandeur of the past in contrast to the triviality of the present. Yet, although Sienkiewicz — like all the novelists of his generation — learned much from Kraszewski's fiction, he very soon went his own creative way and partly rejected him as a literary artist. In his historical fiction, Sienkiewicz challenged Kraszewski on the latter's own ground just as Sienkiewicz's successors were to challenge him on his — style.

Little is known about Sienkiewicz's views on Kraszewski's novels as works of literature. Sienkiewicz left no theoretical views on the "craft of fiction"[3] as post-Jamesian critics understand it. Although fiction flourished exceedingly in the nineteenth century, the "craft of fiction" went almost unnoticed, perhaps because novels were regarded as a form of entertainment rather than an art to be taken seriously. Sienkiewicz's lectures on naturalism in the novel, on the historical novel as a genre, on Zola and other French novelists were concerned

[1] T. S. Eliot, *Selected essays* (New York, 1950), 4.
[2] Włodzimierz Pietrzak, "Drogi polskiej powieści XIX w.", *Twórczość*, II, No. 7-8 (1946), 251.
[3] Janina Kulczycka-Saloni, "Henryk Sienkiewicz krytyk i teoretyk literatury", *Pamiętnik literacki*, XLVII, No. 4 (1956), 389-444, provides a summary of Sienkiewicz's views on literature.

with other aspects. Although Korzeniowski had been concerned with form and structure in his novels, the first Polish novelist to codify a theory of the art was Prus in the 1880s.[4] Still, Sienkiewicz's novels attest to his understanding of the critical problems and the resources of technique in novel-writing, and will be offered as evidence here.

Sienkiewicz published several reviews of novels by Kraszewski in the 1880s. Their tone is respectful, as befitted Kraszewski's eminence as the best-known Polish novelist then writing and the only Polish novelist before Sienkiewicz himself to attain an international reputation. However, these reviews are for the most part descriptive rather than analytical. One exception is the review of Kraszewski's *Krzyżacy* ("The Knights of the Cross"), which Sienkiewicz published in 1881.[5] Here Sienkiewicz devoted most space to commenting on a feature of Kraszewski's approach to the writing of historical novels. This feature was one which Sienkiewicz rejected when he began his *Ogniem i mieczem* ("By Fire and Sword"), in the following year. It was concerned with Kraszewski's tendency to emphasize the historical and factual elements in his novels at the expense of "fantazja" — the imaginative aspect. Sienkiewicz pointed out that Kraszewski's fictitious characters in *Krzyżacy* were "merely ephemeral and sketched shadows", and that the novel proper "meanders round the historical fabric as timidly as a small ivy-branch about the huge tower of some mediæval building." Sienkiewicz added that "history is not the background against which the author depicts their fate" (that of the novel's fictitious protagnosts) "but it is they who are merely appended, in the spirit of the requirements of fiction, to events vastly more huge than they."[6] Professor Krzyżanowski summed up this view when he declared that to Kraszewski "the principles of historical fidelity were so clear that the division between a work of art and a work of scholarship was, in fact, obliterated".[7] Kraszewski consistently put the apparatus of fiction at the service of history.

The problem of the relative importance of fiction and fact in the historical novel preoccupied nearly all nineteenth-century practitioners of the genre.[8] The complexity of the problem can hardly be underestimated: it ranged from stylization of language and the authenticity of setting to questions of historical "truth" and "reality". Adam Mickiewicz had been confronted with a similar

4 David J. Welsh, "Realism in Prus' novel *Lalka*", *Polish Review*, VIII, No. 4 (1963), 33-38, examines some aspects of Prus' literary theories.

5 Reprinted in Henryk Sienkiewicz, *Dzieła* (Warsaw, 1949-55), Vols. XLVI, LI-LII.

6 Henryk Sienkiewicz, *op. cit.*, LII, 249-50.

7 Julian Krzyżanowski, "J. I. Kraszewskiego *Historia w powieści*", reprinted in Wincenty Danek, ed., *Józef Ignacy Kraszewski* (Warsaw, 1962), p. 279.

8 Maria Jasińska, "Autentyzm i literackość a wiedza powieściowego narratora", *Pamiętnik literacki*, LIV, No. 1 (1963), 1, n. 1, and 4, n. 9.

literary problem (in a different genre) when he declared that without his predecessors Karpiński and Niemcewicz, he would never have composed ballads. Sienkiewicz might well have said this about Kraszewski and his own historical novels. In both cases, the writer was faced with the necessity of solving a specific literary problem which his predecessors had solved in their own way, to their own satisfaction, and to that of their contemporaries. But neither Mickiewicz nor Sienkiewicz found their predecessors' solutions adequate: both set about providing their own solutions, and did so primarily by employing certain literary techniques.

The points of relationship between a number of Kraszewski's novels and those of Sienkiewicz were noticed by their contemporaries, including Chmielowski and Matuszewski.[9] These parallelisms occur in "language, characters, subjects and situations, trends and motifs",[10] and have been referred to in general terms by many critics up to the present.[11] We have been promised that "research now in progress will facilitate the publication of evidence based on specific literary facts, to show how much Sienkiewicz owed to the author of *Stara baśń*".[12]

These resemblances are clearest in Kraszewski's 4-volume novel *Kordecki* (1852) and the third volume of Sienkiewicz's *Potop* ("The Deluge", 1884).[13] Both novels were based on at least some of the same sources, both depict the same historical events (the siege of Jasna Góra monastery at Częstochowa by the Swedes in 1655), and both introduce many of the same historical personages. Yet Sienkiewicz's achievement in *Potop* has so far overshadowed that of Kraszewski in *Kordecki* that the latter novel has not been reprinted since 1875, despite the popularity of many of his other novels. Admittedly, Kraszewski himself was aware the novel had faults: he was a penetrating critic of his own work as well as of other people's. But like Scott, he believed that it was better to write much and copiously than to polish or rewrite for the sake of art. The main reason for the neglect of *Kordecki*, however, is to be found in the novelistic techniques used by the respective novelists.

The differences in techniques have received little attention hitherto.[14] There are, after all, many risks inherent in the examination of a writer's technical skill,

[9] Piotr Chmielowski, *Pisma krytycznoliterackie*, ed. by Henryk Markiewicz (Warsaw, 1961), I, 516-9.

[10] Wincenty Danek, ed., *Józef Ignacy Kraszewski* (Warsaw, 1962), p. 100.

[11] Such as Andrzej Stawar, *Pisarstwo Henryka Sienkiewicza* (Warsaw, 1960), p. 263.

[12] Wincenty Danek, *loc. cit.*

[13] Quotations from *Kordecki* are translated from the Lwów edition of 1875. Those from *Potop*, I-III, are translated from *Dzieła*, XI-XIII (Warsaw, 1949).

[14] But see Stanislaw Burkot, "Kraszewski jako teoretyk powieści", *Ruch literacki*, VI, No. 15 (1962), 284-97; Prof. Cz. Zgorzelski's unpublished paper "Sztuka opowiadania Kraszewskiego", referred to in *Przegląd humanistyczny*, VII (1963), 175-6.

since such an examination involves such ill-defined yet all-inclusive concepts as "narrative", "character" and "background", each of which helps constitute the complete pattern of a novel and contributes greatly to its total effect. The technical examination of a novel cannot itself form a reliable guide to the quality of the novel: great technical accomplishment can coexist with an inferior mind,[15] and it is easy to be beguiled into admiring a writer's technical virtuosity at the cost of other, more significant elements. It tells us nothing about the total meaning of a novel, nor of the source of a writer's powers. It is particularly difficult to make any definitive value judgement on Sienkiewicz's novels, the more so as his fame as a "best-seller" has plagued his reputation for the past sixty years.

Yet a writer's meaning can only be presented by means of his technique. Mark Schorer has suggested that "technique is the only means he (the writer) has of discovering, exploring and developing his subject, of conveying its meaning and finally of evaluating it".[16] And an examination of "the arrangement of words on a page"[17] which this kind of approach implies can at least provide one frame of reference for discerning some of the essential differences between two novelists as alike and yet so unlike technically as Sienkiewicz and Kraszewski.

I

It is significant that *Kordecki* and *Potop* were separated by the three decades 1850-1880. This period was marked by the publication of *Madame Bovary* (1857), which was "the source of every important technical advance" (in fiction) "made since the middle of last century".[18] It produced a "gulf between the novels written before and after it".[19] Other fiction published in these decades included *War and Peace* (1868), Zola's *Rougon-Macquart* series (1871-1878) and Maupassant's *Boule de suif* (1880). These writers headed the new, experimental novelists and their influence made itself felt in matters of technique. Flaubert and his successors explored new, more effective technical means of proceeding in narrative, in portrayal of character and rendering of background, and in style. The period marks the emergence in fiction of artistic self-consciousness.

Their influence on the technical development of the novel in Poland has yet to be estimated. Sienkiewicz, who lived in Paris from April to December 1878, did not care for Zola and his followers, but at the same time he had a wide knowledge of contemporary French literature and his own novels indicate that

[15] Robert Liddell, *A Treatise on the Novel* (London, 1947), p. 22.
[16] Mark Schorer, "Technique as discovery", *Hudson Review*, I (1948), 67.
[17] Martin Turnell, *The Novel in France* (London, 1950), p. 6.
[18] Martin Turnell, "Madame Bovary", *Sewanee Review*, LXV (1957), 532.
[19] *Ibid.*, p. 532.

he absorbed many of its new methods into his own technical equipment. Indeed, it was these methods which enabled Sienkiewicz to achieve his "great leap forward" from his Polish predecessors in language[20] and other aspects of his art, and which contributed to making him a modern novelist. In contrast Kraszewski looked back in his fictional techniques to the "flexible novels of loose structure", which the "eighteenth century had bequeathed to the nineteenth".[21] He used many of the conventions employed by such writers as Bronikowski,[22] Niemcewicz and even Krasicki and the writers of the "Monitor" essays.[23] As his art developed, Kraszewski supplemented these techniques with others learned by his own experience, but whereas Sienkiewicz was indebted to his literary traditions, Kraszewski remained in many ways rooted in literary conventions.

Sienkiewicz's skill in narrative was one of the reasons for the enormous popularity of his novels. It is a skill which involves a great deal more than the ability to string together a number of striking or picturesque incidents. It is a technical skill, in that the novelist is always selecting, using, and manipulating various devices so that his material will exert the maximum effect on his readers' sensibilities. This skill implies, in turn, that the novelist is everywhere exerting control over his material. In *Potop* this control is evident in such matters as the length of chapters; preparation; the placing of incidents in such a way as to emphasize or diminish their significance and to make them contribute to the novel's total effect; and in the "pace" of the narrative, which depends on maintaining proportion between scene, analysis, summary, action, and comment.

Kraszewski's *Kordecki* is 150,000 words long. It is entirely concerned with the two-month siege of Jasna Góra monastery, which Sienkiewicz deals with in 60,000 words (the last half of *Potop*, Volume III). There is no reason, of course, why one novelist should not use twice as many words as another. All the same, we should be justified in expecting the latter writer to prove more economical in his means than the former; and paradoxical though it may seem in view of the total length of the six-volume *Potop* (500,000 words), Sienkiewicz was essentially an economical writer in that selection is everywhere apparent in his fiction. Even the length of chapters in *Potop* III provide evidence of Sienkiewicz's economy and selection. The Jasna Góra siege occupies eight chapters (12-19) which range from 15 to 31 pages in length, the final chapter being the

[20] Julian Krzyżanowski, *W kręgu wielkich realistów* (Cracow, 1962), p. 40.

[21] Bradford A. Booth, "Form and technique in the novel", in Joseph E. Baker, ed., *The Reinterpretation of Victorian literature* (Princeton, 1950), p. 68.

[22] Maria Jasińska, *op. cit.*, p. 16.

[23] Ewa Warzenica, "*Latarnia czarnoksięska* wobec oświeceniowej tradycji literackiej", *Ruch literacki*, No. 6 (1962), pp. 303-4.

longest. Each chapter makes its own contribution to the progress of the narrative: Chapter 12 recounts Kmicic's arrival at the monastery with news of the Swedes' approach, Chapter 13 the first appearance of the Swedes, Chapter 14 the start of the siege, Chapter 15 the foray by Kmicic and Czarniecki, and so on.

In *Kordecki*, the length of the 52 chapters varies from four to 42 pages. As with the number of words a novelist uses, so there is again no reason why a novelist should not make his chapters as long or as short as he chooses, providing some specific artistic end is attained. Thus the unusually brief chapters in Tolstoj's *War and Peace* contribute to the panoramic effect of the novel,[24] while on the other hand the long chapters in Orzeszkowa's *Nad Niemnem* help give her narrative its imperturbable spaciousness. The lack of proportion between chapter lengths in *Kordecki* suggests that Kraszewski was little concerned with the "inner architectonics" of his narrative,[25] and indeed it would be wrong to look for any in the work of a writer who was essentially an improviser. Yet it is an artistic deficiency and contrasts with Sienkiewicz's concern with form in even such an apparently mechanical aspect of narrative as chapter length.

The art of preparation is as indispensable in the novel as it is in playwriting. It is one of the most useful devices a novelist can employ to arouse the reader's interest and curiosity, which the novelist can then proceed to satisfy at his leisure.[26] It is equally useful for concealing the mechanics of his plot. Sienkiewicz's contriving and controlling hand is everywhere apparent, though inconspicuously, in *Potop*, which is partly structured by the use of "preparations". An example is provided by the way in which Sienkiewicz prepares us for and leads up to the arrival of the Swedes at the monastery; another by Sienkiewicz's frequent references to the Swedes' heavy cannon with which Miller hopes to bring the siege to an end. The cannons are introduced (III, 293), the Poles spy on them (300), the Swedes use them against the monastery (302, 305-6), Kmicic volunteers to destroy them (307) and does so in Chapter 17. This feat leads in turn to Kmicic's capture by Miller, followed by his escape. As well as providing opportunities for Sienkiewicz to portray Kmicic's character in action, these cannons also play their role in the novel's plot and enable Sienkiewicz to extricate his hero from the besieged monastery without anticlimax.

Sienkiewicz's "placing" of the Jasna Góra siege in *Potop* also attests to the control he exercised over his material. The two-month period occupies precisely half of Volume III (183 pages in a total of 366 pages), and provides the climax

[24] R. F. Christian, *Tolstoj's War and Peace, a Study* (Oxford, 1962), p. 129.

[25] Joseph Warren Beach, *The Twentieth-century Novel — Studies in Technique* (New York, n.d.), p. 164.

[26] The device is also essential in writing for serial publication. *Cf.* my article "Structure and Serialization in the Novels of Henryk Sienkiewicz", *Polish Review*, IX, No. 3 (1964), 51-52.

to the volume. The siege is thus "placed" at the center of *Potop* as a whole, and is depicted as the turning point of the Swedish wars. By "placing" it in this manner, Sienkiewicz helps emphasize the historical import of the siege and elevates it to almost symbolic stature. At the same time, the siege is an important stage in Kmicic's "pilgrim's progress" as he redeems his betrayal of his own (and Sienkiewicz's) moral code. Moreover, there are copious references back to the siege in the subsequent volumes of *Potop*, all of which underline its significance. By placing the siege in such a way as to emphasize its importance, Sienkiewicz was able to let his narrative speak for itself as much as possible and restrict authorial comment to a minimum.

Kraszewski could not do this. As a chronicler, it was part of his function to comment, analyze, and expatiate on his narrative: "There is nothing more terrible than brutalized soldiery: it is something more cruel than a wild beast ..." (III, 28), "Great, oh great is the power of prayer, for it elevates the soul, purifies us, mysteriously changes us ..." (III, 47) or "How terrible is war when, dragging countless disasters in its wake, it crosses a country with fire and sword ..." (I, 94). Kraszewski never hesitated to intervene in his narrative: "Let us look once again at the Swedish camp in order to learn the reason for Miller's sudden change ..." (IV, 209), to remind us of what has gone before (Kuklinowski "has already been mentioned by us" [III, 44]), and to provide footnotes within the text: "Monks in the seventeenth century were still bound by the main rules of St. Augustine's regulations — nocturnal prayers, hair-shirts, belts, discipline, fasts, confessing the sick, burying the dead ..." (II, 110). The "talkative" author who haunts nineteenth-century fiction was deposed by Flaubert with his comment "One must not bring one's own personality on the scene".[27] Yet authorial comment is a legitimate device in narrative, especially of the chronicle sort. It is not necessarily inferior to more objective, impersonal methods of narrating, despite its tendency to bring the narrative to a halt and weaken illusion. Sienkiewicz was not averse to discreet, unemphatic, often ironical comments which he used to link scenes or to provide expository material in an economical way. The effect of authorial comment depends most of all on the tone in which the comments are made. Kraszewski's are intrusive in their rhetorical floridity, especially to readers whose sensibility has been modified by later authors, including Sienkiewicz himself.

Interventions by the author in fiction were criticized in Poland by Chmielowski in the 1880s.[28] Sienkiewicz did not codify his views on the device, which he

[27] Gustave Flaubert's letter to George Sand (1866) quoted by Miriam Allott, ed., *Novelists on the novel* (New York, 1959), 271.

[28] Maria Żmigrodzka, "Problem narratora w teorii powieści", *Pamiętnik literacki*, LIV (1963), 425.

used freely in his first published novel *Na Marne* (1871). In his later novels (after 1882), the talkative author has almost entirely disappeared, except for an occasional rhetorical question ("What could be more terrible than the news of Jan Kazimierz's abdication?" [III, 264], and "What then was happening to Andrzej and in what way did he succeed in bringing his purpose about?" [III, 317]). But the tone of these questions, like the tone of Sienkiewicz's comments, is unemphatic and easily assimilated.

Kraszewski's narrative proceeds at a slow, relaxed pace. The novel is structured on strictly chronological order, with historical time as an organizing principle. All the important events are dated: "On the morning of November 1, 1655 ...", "On November 3rd ...", "On the Thursday ...", and so on, up to and including Christmas Eve and Day. The dates are often prominently placed at the beginning of chapters, so that they serve to introduce events. Their function was to intensify the impression of authenticity with which Kraszewski was primarily concerned. Like Walter Scott, Kraszewski believed that the interest and value of his novel lay in the accuracy with which it presented historical characters and incidents. The dates contribute to this accuracy by anchoring the narrative firmly in historical time. This is the technique of a chronicler, and accords perfectly with other features of the chronicle which Kraszewski employed.

Fictitious time (like fictitious characters) was more important to Sienkiewicz than historical time. He sometimes used dates, but like his comments they are introduced into the narrative in an unemphatic manner, often combined with natural imagery: "The night came on dark and rainy. It was November 8. An early winter had set in and the first flakes of wet snow, with waves of rain, flew to the earth ..." (III, 208). But Sienkiewicz was more concerned with "fictitious" time, which he could expand or contract in proportion to the importance he wished to accord incidents in his narrative. A night passes in one phrase (203), three days are foreshortened to as many pages (304-6), while the events of a few hours can expand to occupy ten or 20 pages. Kmicic's confrontation of the monks and gentry after his arrival at Jasna Góra extends over pp. 189-99, Kuklinowski's embassy to the monastery (282-91), Kmicic's exploit with the Swedish cannon occupies pp. 317-336. This method of expanding dramatic scenes to their fullest extent (mostly by dialogue) ensures that scenes so treated will produce their maximum effect on the reader. It is a characteristic feature of all great nineteenth-century novelists from Dickens and George Eliot to Tolstoi and Dostoevskij, and has become a commonplace of present-day best-sellers.

Kraszewski also had the power of extending and heightening dramatic scenes in this manner. But it is significant that long scenes between Kordecki and other historical characters, based on documentary sources, are stylistically dead, couched in florid rhetoric and *oratio obliqua* (I, 35). On the other hand, the scene in which the aged Lassota (a fictitious character) spurns Miller's order to drink the health of the Swedish king (I, 101) is one of the most effective incidents in the novel and clearly impressed Sienkiewicz. It foreshadows the banquet scene (*Potop* I, 265-86) during which Janusz Radziwiłł calls upon the assembled gentry to toast Charles Gustav, an act which symbolizes Kmicic's betrayal of the Polish cause.

II

Kraszewski's treatment of his characters was consistent with his chronicle technique. The historical protagonists in *Kordecki* — Kordecki himself, Miller and his aides, Czarniecki and the other Polish gentry — are in the foreground throughout the novel. It is they who provide the novel with its main interest and value. The fictitious characters who take part in a conventional sub-plot are the "ephemeral and sketched shadows" Sienkiewicz was later to refer to in the *Krzyżacy* review. They do not appear until near the end of Volume I, play no part in the historical events and are neglected for long sections of the narrative (the "heroine" is kidnapped half-way through Volume III and is not seen again until the end of Volume IV). The sub-plot was of Kraszewski's own contrivance and depends for its effect upon the novelist suppressing information about the characters' relationships. Its function was primarily to entertain, while the historical characters and events were intended to provide edifying examples. In this respect, *Kordecki* combined the useful and the entertaining elements characteristic of much fiction of the eighteenth and early nineteenth centuries, when the novelist was a teacher as well as a purveyor of rational entertainment with a high moral tone.

Sienkiewicz focussed consistently upon his fictitious character Andrzej Kmicic, who is the center of attention throughout *Potop*. The historical personages who occupied the foreground in *Kordecki* are here presented by Sienkiewicz almost exclusively in terms of their relations to Kmicic, and we see them as Kmicic sees them. They become a background against which Sienkiewicz displays Kmicic's character in action. In the same way, the Swedish wars became a background to Kmicic's *Bildungsroman* as he seeks fulfilment in the conflict between himself and the world.

By depicting the historical characters present at the siege as they appear to Kmicic, Sienkiewicz was able to avoid the pitfalls of attempting to enter convincingly into the consciousness of a "real" personage. The device of portraying a novel's characters through the eyes of another is one of the most effective means available to a novelist of making his readers "see", and — like Joseph

Conrad — Sienkiewicz was always concerned with this. Thus we are made to share Kmicic's perception of Kordecki: when Kmicic is brought into the prior's presence, he sees "a man of mature years, already approaching evening. He had a very tranquil face. A copious beard framed his features and he had blue, calm, and penetrating eyes. In his white habit he had the look, simply, of a saint ..." (III, 189). He is not named as Kordecki until the audience is half over, because Kmicic does not know his name either. Sienkiewicz provides only those details of Kordecki's appearance which Kmicic would notice.

Kmicic himself was introduced through the eyes of another character in the first volume of *Potop*: the heroine Aleksandra Billewiczówna is anticipating his arrival:

> Then, in the doorway, appeared a towering figure in a fur-lined coat and cap. The young man stepped into the chamber and realizing he was in the servants' anteroom, asked in a sonorous voice:
> "Hey! Where's your mistress?"
> "I am here", Billewiczówna answered in a fairly assured voice.
> Hearing this, the newcomer took off his cap, threw it to the floor and said with a bow:
> "I am Andrzej Kmicic".
> Pani Aleksandra's eyes rested for an instant on Kmicic's face, then were again cast to the floor: in that moment, however, the girl had been able to observe a crop of hair flaxen as rye, much shaven, a swarthy complexion, grey eyes gazing keenly, a dark moustache, and young aquiline face. (I, 13)

This device, used sporadically by many earlier novelists, was first employed in a consistent and deliberate manner by Flaubert.[29] Sienkiewicz also used it consistently to intensify the visual element in introducing characters and heightening moments of drama in the narrative. Thus Bogusław Radziwiłł is presented as "seen" by Kmicic (II, 165), as is Wrzeszczowicz (III, 172). Kmicic's appearance at the Lubicz manor-house in response to Wołodyjowski's challenge is presented as "seen" by those present (I, 171): Kiemlicz and his son are "seen" by Kmicic (III, 22), and Kmicic is described as suddenly observed by De Fossis when he enters the latter's tent (III, 252). Sometimes these highly visualized scenes are emphasized by lighting effects, as when Józwo thrusts a burning brand before Kmicic's face so that his enemies suddenly recognize him by its light (III, 63), and Kordecki on the monastery walls is illuminated by an ostensory (III, 225). Similarly, Sienkiewicz often describes aural effects in terms of the hearer (Kmicic "hearing" Kordecki's sermon, III, 189).

Kraszewski presents his characters in a manner much favored by his predecessors, from Balzac and Scott to Niemcewicz.[30] The introduction of a

[29] Anna Granville Hatcher, "*Voir* as a modern novelistic device", *Philological Quarterly*, XXIII (1944), 362.

[30] Cf. the "portraits" of Zygmunt August, Bona, Izabella, Zygmunt, and Barbara in J. U. Niemcewicz, *Jan z Tęczyna* (Wrocław, 1954), pp. 80-82.

character provides the novelist with an opportunity to enumerate facts and details of the character's appearance, costume, past history, qualities, and vices. The information is provided in the form of non-dramatic analysis, which the novelist interprets. Like his predecessors, Kraszewski believed a novelist's place was between his characters and his readers. It was the novelist's function (like that of a chronicler) to describe, analyze, and comment on his protagonists, whether historical or fictitious, just as he described, analyzed, and commented on incidents and backgrounds.

Kraszewski's portrait of Kordecki is designed to predispose us in the latter's favor:

> Kordecki was a man still of middle-age, of unexceptional height and with a face not at all heroic; his features were characterized only by a powerful soul, an inflexible nature and lucid intellect; benevolent mildness joined in them fortitude and constancy. His grey eyes gazed keenly, directly at everyone, not yielding before any, nor fearing inquisitorial scrutiny, themselves penetrating to the soul's depths. (I, 26)

Kraszewski also provides a summary of Kordecki's parentage and past life, and he comments: "The seventeenth century in our country was still an age of sincere and deep faith, but even in it there were few men like Kordecki. ..." (I, 27). In the same way Kraszewski summarizes and comments on the Swedish general Miller as "a soldier and a boor, laughing at the mischief and crimes of his soldiery, hanging and killing, slaughtering, marauding and feeling neither qualms of conscience nor disgust at their trade. ... The greater part of the Swedes crossed Poland in this manner, and to this day their old castles attest how furiously, irresponsibly, and mercilessly they marauded the country." (I, 84-5).

As an item of novelistic technique, the method harks back to the "portraits" in eighteenth-century novels and satires, which in turn derived from the "character" writers of the seventeenth century, such as La Bruyère.[31] It was a legitimate method, and could be economical and effective when skilfully used. Sienkiewicz did not hesitate to incorporate it in his battery of technical devices, and presents the renegade Kuklinowski in *Potop* III in precisely this manner. However, Sienkiewicz makes the "block characterization" of Kuklinowski serve a double purpose; in addition to presenting Kuklinowski as an individual Sienkiewicz also depicts him as representative of all the Polish renegades. Kuklinowski

> was one of those soldiers in whom the conscience had been seared as the result of a debauched life and continued campaigning. Many such prowled about Germany and

[31] David Nicol Smith, *Characters from the Histories and Memoirs of the Seventeenth Century* (Oxford, 1908), and Benjamin Boyce, *The Theophrastan Character in England* (Cambridge, Mass. 1947), provide many instances.

Poland after the Thirty Years' war. ... Their homeland, faith, and all that was holy were indifferent to them. They knew only soldiering, in it they sought consolation, debauchery, profit and forgetfulness. ... Such was Kuklinowski. (III, 282-3)

Elsewhere, Sienkiewicz did not hesitate to summarize and comment in his portrayal of episodic characters: Sadowski "was a bold and impetuous soldier and, since he knew his own worth, took offence easily" (III, 222) and Sienkiewicz then depicts him doing so. The Swedish general Miller "disliked the Duke of Hesse, but like many men of low origins, was impressed by his cold, civil, and at the same time contemptuous manner of speaking, proper to men of high quality" (III, 280). But the comments are always brief and economical, their tone epigrammatic and often ironical, so that they contribute to the narrative instead of holding it up.

III

Elizabeth Bowen has suggested that "scene is only justified in the novel where it can be shown, or at least felt, to act upon action or character, where it has dramatic use".[32] Like all great novelists, Sienkiewicz was fully aware of this function of setting; even in his highly pictorial *Quo vadis?*, descriptive passages are not introduced for their own sake, but to contribute to character and action.

Sienkiewicz frequently depicts settings as "seen" by one or more of his characters, who thus become the medium by which the setting is conveyed to the reader. Background is presented in terms of the effect it has on the novel's protagonists. The Jasna Góra monastery is first shown as "seen" by Kmicic and his followers, who

stopped singing and looked in that direction. Meanwhile the sky grew lighter, changed from pale to green and gold, and a point on the horizon began glittering, so that they blinked at its glow. Soroka said: "Is it a miracle, or what? The west is over there, yet that is like the sun-rise".

And in fact the light increased before their eyes, from a mere dot it became a globe, from a small globe a large one — at a distance you would have said that someone had suspended a huge star over the earth ...

Kmicic and his men looked in amazement at this luminous sight, quivering and gleaming, not realizing what it was they saw. (III, 183).

As they drew closer,

the church, monastery, and surrounding walls became clearer, grew more splendid, more immense. Finally they perceived the town as well, in the distance, and whole rows of houses and huts at its foot, looking as small as birds' nests compared to the immensity of the church. (III, 184)

[32] Elizabeth Bowen, *Collected Impressions* (London, 1950), p. 254.

In these passages, Sienkiewicz presents the scene in terms of perspective, a favorite device of seventeenth-century Baroque poets such as Twardowski, who depicts Minerva's temple as it appears to Pasqualina, with its "glittering dome" visible in the distance.[33]

Jasna Góra is also shown as it appears to the Swedes camped outside:

A small rain began drizzling in the form of a cold and penetrating mist, which in places formed into pillars, columns, and bridges, reddening from fire.

Amidst these fantastic pillars and arcades was visible from time to time the looming outlines of the monastery, which changed shape before their eyes: now it looked loftier than usual, then as if sinking into an abyss. From the trenches up to the very walls gloomy arches and corridors formed from the mist and obscurity. ... Sometimes the air above the monastery became clear, as if illuminated by lightning. Then the high walls and towers were outlined brilliantly, then again extinguished. ... The soldiers began watching with sombre alarm and superstition. (III, 242)

and

A white storm veiled the topmost turret of Jasna Góra and ought, by the nature of things, to have covered the church; but by a most extraordinary phenomenon of Nature, the church and tower rose not only above the cliff but high above the mist ... as if it had left its foundations and was suspended in the sky. ... The shouts of soldiers attested they had seen this. (III, 258)

Landscape apparently in motion also appears in poetry of the Polish Baroque: Zimorovic's description of Lwów describes how "its tall walls / Raise their towers from where the ground crawls / Up into brightness, colored bright, / With their tops ... they touch the Heavens".[34]

The interior of the monastery, as seen by Kmicic, is likewise described in terms of light, shade and color also in motion:

A red darkness prevailed in the chapel, not completely dispelled by candle flames burning on the altar. Colored lights also fell through the windowpanes and all these — red, purple, golden, fiery — quivered on the walls, shifted over the effigies, the arches, penetrated into obscure depths to bring out vague shapes apparently plunged in slumber. ... Everything was half-visible, half-veiled, unearthly. (III, 187)

We perceive this Baroque interior through the consciousness of the simple, unreflective, religious Kmicic: the details are selected in order to heighten Kmicic's mood, to illustrate his character and emphasize the evocative power of the scene.

Kraszewski's settings are the utilitarian scene-settings of much eighteenth-century fiction, and are as factual as his presentation of character. He uses the

[33] Samuel Twardowski, *Nadobna Paskwalina* (1655), III, 628-30, 637-9.

[34] Quoted from Jerzy Peterkiewicz and Burns Singer, *Five Centuries of Polish poetry* (London, 1960), p. 50.

description of Jasna Góra which opens *Kordecki* to establish his own omniscience: here the novelist becomes a guide, pointing out and commenting on the scene. It is significant that Kraszewski first surveys the monastery as it looks "today" and uses the present tense in his description: Balzac (*Le Père Goriot*) and Stendhal (*Le Rouge et le noir*) also used the present tense in the descriptive passages which open these novels. By affirming that the geographical places or buildings against which the novel is to be set are "still there", the novelist guarantees the authenticity or at least the objectivity of the events which "once" occurred there. Kraszewski then proceeds, by a brief comment, into the historical aspect of his scene ("How can we not but say a few words of the history of this place ...?").

We can respect the painstaking erudition of these and other historical passages, but they are not an adequate substitute for the highly imagined re-creation of settings in Sienkiewicz's novel. The descriptive passages in *Potop* — unlike the historical digressions and enumerations of buildings in *Kordecki* — cannot be detached from the narrative. They contribute more than mere setting: Kmicic's view of the monastery and its interior illustrates his character, just as Miller's superstitious awe when he sees the monastery illuminated and apparently in motion contributes to our knowledge of his. Instead of enumerating, summarizing, and generalizing, Sienkiewicz gives his scenes emotional significance by dramatizing them.

Were it not for the influence which Kraszewski's *Kordecki* and other early novels exerted upon Sienkiewicz, they would have little more than an antiquarian interest. Later in his career Kraszewski's technical equipment increased, enabling him to compose such novels as *Cosel* and *Brühl* (1873-4) which have an established place in the Polish "house of fiction". Yet Kraszewski's skill as biographer and historian, his intellectual qualities, his high moral purpose, and his industry suggest that these are not necessarily the aptitudes essential to a novelist. He lacked those insights which, no matter what techniques are used to communicate them to the reader, make the difference between a competent storyteller and a great novelist possessed of a powerful imagination.

THE 1073 *IZBORNIK*: THE MANUSCRIPT AND ITS SOURCES

ROBERT H. WHITMAN

Indiana University

In tracing the fitful history of the formation of the Russian literary language, we have begun to appreciate the important role played by the special "literary" traditions which are collectively designated Church Slavonic (CS). The evolving East Slavic vernaculars provide the continuous backdrop against which this history must be followed. But it was the various normative, often foreign-inspired scribal practises which were cultivated in monastery scriptoria and the like and the scribes' sense of the peculiar status and integrity of such literary practises, together with the growing fund of MSS., that served as the abiding catalyst in the long process.

A better understanding of the events depends now on a more systematic and detailed examination of the MSS. which embody such relatively strict norms during different periods. Large MSS. provide the most reliable samples of what a given scribe or "school", in spite of sometimes frequent irregularities, regarded as correct, standard usage. Especially important are individual Russian CS MSS. or families of them whose roots can be traced to the period just before and after the official adoption of Christianity by Russia in 989 and, even more distantly, to the literary tradition which developed during the flowering of the Bulgarian state of Simeon around the year 900. The pattern of the CS tradition, if not necessarily the details, was set in the earliest period of transmission from the South and West.

The so-called *Svjatoslav Izbornik of 1073* (hereafter, 1073) is one of the few rich sources surviving from this era. Roughly a century and a half after its discovery, a start has recently been made on thorough investigations of its lexicon and morphology.[1] The purpose of the present study is to assemble the scattered earlier scholarship on the MS. and to comment on a number of textual questions.

[1] G. H. Worth's study of Greek-influenced vocabulary, to appear in a subsequent issue of this series, and my dissertation, "The Morphology of the *Svjatoslav Izbornik of 1073*"(Harvard University, July 1963).

Most of 1073's over 260 pages are devoted to exegetic patristic prose.[2] The central and largest portion (27b20-223b28) is occupied by a part of the famous "Questions and Answers of Anastasius of Sinai".[3] This is preceded by a portrait of Svjatoslav II, Grand Prince of Kiev, for whom the text was ostensibly copied, and of his family (1 recto), a dedicatory poem to Svjatoslav (p. 2), a subtitle (4 recto top),[4] selections from a number of *slovo*, largely from the Greek Church Fathers (4 recto-23b14) and a survey of six major Church Councils (23b15-27b19).

Following the Anastasius material are additional patristic selections (223b29-263c22; including Choeroboscus' treatise on rhetorical figures[5] and the "Dialogue between Gregory the Theologian and Basil"), a concluding statement (263c23-25), a statement of the date and authorship (263c26-29),[6] a slightly altered repetition of the dedicatory poem (263d1-264a24),[7] and a briefly annotated list of Greek and Latin sovereigns "*otъ Avъgusta daže i do Konstantina i Zoja* ..." (264a25-266b9). An index is inserted roughly in the middle of the text (123a-127b) to cover the second half.

Several other manuscripts containing substantially the same material have been preserved. L. Masing, in an early article appearing in two parts,[8] has done a comparative tabulation of the ordering and selection of the material with respect to a possible Greek source and to the original Slavic translation.

On the basis of this work, Masing has been able to correct the page order of the present binding of 1073 in several places. The biggest disturbance exists near the inserted index at the end of the first part. The correct sequence is as follows: 117, 149-152, 122-130, lacuna, 131-148, 118-121, 153 (Masing, pp. 95

[2] 'Page' refers to both sides of a leaf. 'a' and 'b' refer to the two columns recto, 'c' and 'd' to the two columns verso. These are the designations used in Šaxmatov's errata. See below. The number after a letter refers to the line.

[3] The heading reads: "Anastasiejevi otъvěti protivu nanesenyimъ jemu otъvětomъ ně otъ kakyixъ pravověrьnyixъ o različьnyixъ glaviznaxъ" (27b).

[4] Sъborъ otъ mnogъ o[tь]cь, tьlkovanija
o n[era]zumьn[yi]xъ slovesьxъ
[vъ e]üa[g]gelii, i vъ ap[osto]lě, i vъ
in[ěxъ] knigaxъ, vъkratьcě sъloženy
na pamjatь i na gotovъ otъvětъ.

[5] An extensive study of this selection has been made by Justinia Besharov in *Imagery of the Igor' tale in the light of Byzantino-Slavic poetic theory* (Leiden, 1956).

[6] "Vъ lěto 1073 napisa Ioannъ diakъ izborьnikъ sь velikuumu knjazju S[vja]toslavu."

[7] Rajko Nahtigal cites earlier publications and reconstructions and provides his own version of the original poem in Part II (pp. 123-135) of "Rekonstrukcija treh starocerkvenoslovanskih izvirnih pesnitev", *Razprave Akademije znanosti in umetnosti*, I (Ljubljana, 1943), 3-116.

[8] "Studien zur Kenntniss des Izbornik Svjatoslava vom Jahre 1073 nebst Bemerkungen zu den jüngeren Handschriften", *Archiv für Slavische Philologie*, VIII (1885), 357-395; IX (1886), 77-112.

[and first fn.]-100). My references are based on the present numbering of the binding.

There are six manuscripts in all, of which 1073 is the oldest. At the time of its last complete catalogue description,[9] it was registered as No. 161 in the holdings of the Synodal Library in Moscow. It is now in the State Historical Museum in Moscow. The circumstances of its discovery in 1817 and of its subsequent wandering are described by Elpidifor Barsov in his introduction to a partial edition of the MS.[10] We shall refer to this introduction again below.

Two other MSS., which Masing designates S^2 (No. 162 of the Synodal Library; described by Gorskij et al., pp. 406-409) and B, date from 1403 and 1519, respectively, and derive from 1073 itself. S^2 is a direct copy with some few omissions (Masing, p. 360). B corresponds very closely to S^2. However, the nature of one of the discrepancies suggests that at least one MS. lies between S^2 and B (Masing, pp. 367-369). Both MSS. (but particularly S^2) are useful for reconstructing lost or damaged pages and for dating additions and alterations in 1073 against the 1403 date (Masing, pp. 269-270).

Another copy of 1073, R^2, was made in 1818 and 1819 on the commission of Count Rumjancev, who had also commissioned the expedition which discovered the MS. This copy, which was planned for publication (Barsov, p. iii), was made by a certain Collegiate Secretary, Ratšin. Barsov quotes Ratšin's own description of the procedures followed (pp. v-vi). R^2 was filed as No. 356 in the Rumjancev collection and was described by Vostokov.[11] It contains the pages which are missing in S^2.

The most important MS. for establishing an approximate genealogy for 1073 is the one designated K by Masing which was discovered by Ševyrev in 1847. The dating of K is uncertain. However, Masing thinks it may date from about 1500 (pp. 373-374). There are extensive correspondences with 1073 which, in spite of many divergences, indicate that the two MSS. derive from a "not too distant common source" (Masing, pp. 388-389). The title (*Sъborъ ot*[*ъ*] *mnogъ o*[*t*]*cь tolkovanija...*", Masing, p. 377) is almost identical to that of 1073. Of particular interest is the dedicatory poem to Simeon of Bulgaria (Masing, pp. 374-375), whose name is appropriately replaced in 1073 by that of Svjatoslav. Otherwise the wording is substantially the same (See nn. 6 and 7). Simeon reigned from 893 to 927, and it is fairly clear that the mention

[9] A. Gorskij and K. Nevostruev, eds., *Opisanie slavjanskix rukopisej Moskovskoj Sinodal'noj biblioteki*, II, No. 2 (Moscow, 1859), 365-405.

[10] *Izbornik Velikago Knjazja Svjatoslava Jaroslaviča 1073 goda* (= *Čtenija v Imperatosrkom Obščestve Istorii i Drevnostej Rossijskix pri Moskovskom Universitete*, Kniga četvertaja, Oct.-Dec. 1882; appearing in Moscow, 1884, dated "1883").

[11] Aleksandr Vostokov, *Opisanie russkix i slovenskix rukopisej Rumjancovskago muzeuma* (Saint Petersburg, 1842), pp. 499-506.

of his name establishes this period as the terminus for the original translation from the Greek. We shall return to the question of this lost original below.

Masing's R^1, which was filed as No. 6 in the Rumjancev collection and also described by Vostokov (pp. 9-10), consists almost exclusively of the Anastasius material and is more distantly related to 1073 than is K. In general, the contents match the corresponding passages of 1073 rather closely. As Masing points out (p. 357), the "Questions" of Anastasius is an established Greek patristic text which exists independently from the presumed Greek *izbornik* or miscellany. Nevertheless, R^1 appears to derive not from the Greek Anastasius, but from the *izbornik*. First, R^1 contains several passages which are not properly part of the "Questions" and which it shares with 1073 and K (Masing, pp. 109-110, 392). Second, the *izbornik* contains only selections from the Anastasius text. In R^1 these are almost the same and in the same order as in the other (explicitly "miscellaneous") Slavic MSS. The last point deserves particular note, because after Question 59 the sequence of questions in all the Slavonic MSS. (which fall into at least two distinct lineages) has many omissions and rearrangements with respect to the numbering in Migne's edition of the Greek (Vol. 89; Masing, p. 103f.). It seems reasonably certain that the original Slavic translation was of similar shape.

There is little question that a Greek *izbornik* of comparable make-up did exist. Masing cites Montfaucon's ancient description[12] of MS. No. 120 ("formerly 119") from the Coislin Library (p. 359). A modern description of the collection, including the *izbornik*, has been made by Robert Devreesse.[13] Devreesse also has described another MS. of similar composition from the Vatican and bearing the number 423.[14] Both Greek texts show the following sequence for the Anastasius Questions: 1-59, 142-151, 60-74, 152-154. The only discrepancy, according to Masing's tabulation (p. 107), is that all the Slavic texts also have Question 128 inserted after Question 70. The absence of 49-52 is not accompanied by a smooth break (p. 102 and second fn.) and must be a secondary phenomenon.

Most of the other divergences between Migne and the Slavic texts are minor. Usually a sub-selection is either missing or shorter in Migne. However, Migne is based on a somewhat corrupt text.

On the basis of his inspection of a copy of the Coislin MS., Vostokov believed that the Slavic line of manuscripts must originally have come from another

[12] Bernard de Montfaucon, *Bibliotheca Coisliniana, olim Sequeriana* (Paris, 1715), pp. 192-195.

[13] *Catalogue des Manuscrits Grecs, II, Le Fond Coislin* (Paris: Bibliothèque Nationale, Département des Manuscrits, 1945), pp. 109-111.

[14] *Codices Vaticani Graeci*, Tomus II (= *Codices* 330-603; The Vatican, 1937), 138-141.

Greek manuscript (p. 500). In the absence of a reasoned and documented argument to support that contention, I regard the judgment as somewhat premature. The contents parallel each other so closely that we would have to have very strong proof that a particular discrepancy could not have arisen in the single Slavic source for all the manuscripts. Vostokov's apparently naïve approach to the philological complexities of the relationships among MSS. requires, I gather, that a meaningful derivation of MSS. be based solely on virtual identity.

If Masing, for his part, based his support of Vostokov's position (p. 359) only on the Montfaucon description, his observations should be disregarded. Devreesse's descriptions are more detailed and dependable, and they correct a number of misleading features of Montfaucon. For example, Montfaucon (p. 193) jumps from a selection from Saint Augustine (folio 213) directly to one from Hesychius (folio 214). Devreesse, in describing the same place, notes that two folia are lost. He also points out that the beginning of a selection from Eusebius ("ex chronicis") is damaged (p. 110). The Vatican MS. supplies what was contained in the lost Coislin folia, namely, the selection "de apparitione Domini, ex Constitutionibus Apostolicis" (p. 140). Working exclusively with Montfaucon, Masing could readily have concluded that two selections in 1073 (*O obavl'enii gospodьn'i otъ ap[osto]lьskyxъ zapovědii,* 247c17 ff.; *Eüsevijevo otъ lětn'iixъ,* 247d28 ff.) represent corruptions, and his impression of the differences between 1073 and the Greek would have been correspondingly distorted.

The single major discrepancy between 1073 and the Greek MSS. also separates 1073 and K from R[1]. Toward the end of the so-called first part of those two MSS. occur four short patristic selections which have a form suitable for the Anastasius "Questions" as a whole, but which are thematically unrelated to either the preceding or the following Question and appear to be later insertions. Although 1073 numbers them, K does not (Masing, p. 95). Furthermore, they are not listed in the index to the first part of K. (There is no index for the first part of 1073.)

Unfortunately, the method of simple juxtaposition of textual differences inherent in Masing's tabular presentation imposes a false perspective and obscures the relationship between the insertions and other disturbances in the same region of text. Because of it, Masing is forced to view everything indiscrimately in the same plane. But in order to discuss textual families properly, it is essential that we try to display the individual developmental strata underlying the present condition of the passage in all the MSS. The serious consequence of Masing's approach is that, like too many investigators, he tends to exaggerate the importance of the single "original" Greek source as a point of reference, to magnify deviations from it and, in general, to ignore the multiple possibilities for disturbances and for later accretions — from both Slavic and other Greek sources. Before turning to the problem of the source of the extra selections under consideration, let us review the events which clearly seem to have preceded their intrusion. These are important since the three major MSS. (1073, K and R¹) apparently shared an identical fate in their fundamental layers. The differences between them, as a group, and the Greek are, furthermore, much simpler and less significant than Masing's table in-indicates.

The basic step, found in all the MSS., was probably a single break or division in the binding of the folia. It would not be surprising if such a large MS. were divided into two parts, and it is interesting to note that the Vatican MS., No. 423, happens to be broken at Question 24, very shortly after the point where the Slavic MSS. indicate a break (Devreesse, 1937, p. 138). Subsequently two sections of text were broken off from one or each exposed end and simply reversed. If we number the six words or phrases adjoining the potential breaks in undisturbed sequence, the order exhibited by the surviving MSS. in 1/4, 5/2, 3/6 — with slash indicating the present splices. These numbers identify the units in the further discussion below.

Of the three junctures thus created, only the first is still shown in 1073, which, according to Masing's estimate (pp. 96-97), is missing eight folia after page 130. The preserved juncture occurs after *sutь* in 151b20. All three are preserved in R¹: 122 recto 21 after *sutь*, 125 verso 7 after *prěobidemomъ*, 128 verso 5 after *těxь*.

As a curious coincidence, the superficial similarity in CS between 1 and 3 and between 4 and 6 seems to have facilitated the transposition. The first splice is bounded by (1) *o*[*t*] *nixъže sutь* / (4) *jazvьni byv'še* in R¹ (*otъ n'ixъže sutь* / *jazveni byvъše* in 1073), the third splice by (3) *o*[*t*] *těxь* / (6) *javitisę*. The correct Greek — with equivalent units divided and numbered — is as follows: (1) ἐξ ὧν εἶσιν / (2)'Ιαννης, και 'Ιαμβρος; (3) ὑπ' αὑτων / (4) τραυματιαι, γενομενοι; (5) εὐκαταφροντος / (6) φανειται. The two remaining Greek units, 2 and 5, are matched in the Slavic at the middle splice: (5) *prěobidemomъ* / (2) *jakože anьni i amvri*. There is no Greek equivalent for *jakože*; it was apparently introduced for the sake of continuity.

The condition of R¹ (with no disturbances other than the transposition) establishes quite firmly that the division of 1073 and K into two parts was not a feature either of the original Greek or the original Slavic translation. The second step appears to have been the compilation for a MS. underlying only 1073 and K of an index which now more or less formalized the division. In 1073 it runs from 123a1 to 127b29. Preceding the index is an incomplete column (122d) which is partially filled by a sketch and by six lines of unrelated text which Vostokov (p. 502) assigns to the 12th or 13th century. Immediately following the index was a blank side which was also filled by a somewhat later hand.[15] 128 recto and verso are completely occupied by illustrations.

The reasons for the appearance of the index are not entirely clear. Nor is it entirely clear why the index should have begun only at this point, a little beyond the middle of the text. A likely guess is that a MS. underlying 1073 and K had by this time acquired at least a partial ordered list of the contents without numerical coordination between the selections and the list, and that the formal index was modeled on it and intended as a complement to it.

The index and the completely coordinated numbering of selections in both 1073 and K show that the two transposed stretches of text were included with the second part. As further indication, we have the fact that the repetition of the whole *izbornik*'s title in K, although subsequently separated from the body of the second part by the index and by one of the extra selections (see below), concludes with run-on short titles for the first two selections of the second part (*Theodoritovo* ..., *Kurilovo* ...; Masing, p. 380). Presumably these special identifications were felt to be necessary since major sub-sections within the Anastasius portion of the *izbornik* do not begin with what was now found there; the strict form for each Question is, in this order, the question itself (ordinarily, brief), Anastasius' own (fairly lengthy) response and then only a selection of relevant short passages from the Bible or other patristic texts, such as the two just cited.

The ground was now prepared for the third step, the inclusion of the extra selections. Again we cannot be completely certain about the reasons, but the following seems likely. The transposition principally involved Question 20. The passage bounded by Units 2 and 3 runs from near the end of the penultimate paragraph of Anastasius' answer to Question 20 to the middle of the sixth sub-selection of the same Question included in the Slavic translation. The one bounded by 4 and 5 runs from the latter point to roughtly the middle of the second paragraph of Anastasius' answer to Question 21. However, the interruptions in both answers were effectively concealed, because the transposition simultaneously moved substitute endings into position. The answer to Question 20 now ended with the last half of its sixth sub-selection, the answer to Question 21 with the last paragraph and a fraction of another of the answer to Question 20. Question 20 in particular gives the appearance of completeness; it would have taken an exceptionally attentive scribe armed with the Greek original or a copy to detect the disturbance.

Nevertheless, if it was regarded as unusual to begin the second half of the

[15] Gorskij estimates roughly the 13th century (p. 374), Nikolaj N. Durnovo the beginning of the 15th century ("Russkie rukopisi XI i XII vv., kak pamjatniki staroslavjanskogo jazyka", *Južnoslovenski filolog*, IV [1924], 77).

izbornik with minor sub-selections from a Question, it would be equally undesirable to end the first half simply with Anastasius' answer, without any of the normal sub-selections. Since that was precisely where the answer to Question 20 now stood, it is easy to see why the scribe of a MS. underlying 1073 and K should have borrowed suitable sub-selections of the type we find in these two MSS. from another text.

Finally, 1073, unlike K, exhibits an additional feature in the same area which could be a pure inadvertence on the part of the scribe or possibly a further, tentative solution to the problem of the unusual position of the answer to Question 20. This is the anticipation at the end of the first half of the *izbornik* but before the four borrowed selections of all of the first (*Theodoritovo* ...) and part of the second selection (*Kurilovo* ...) of the second half. The duplications are found at 151b23-151d13 and 151d14-20, respectively. Both explanations are to some extent complicated by the fact of the four borrowed selections. If they were already present somewhere near the end of the first half of the *izbornik*, it is not clear why the scribe should have jumped ahead to the first two selections of the second half. However, as we shall see in a moment, the position of the borrowed selections probably was not completely fixed as yet.

Scribal carelessness is possibly the more complicated explanation, since it requires assuming a different placement of the borrowed selections than we find in 1073. But given that much, a scribal oversight would have been quite easy, particularly if the individual selections were still not numbered in accordance with the new index. At the same time, the existence of such numbers might have been the very reason why the scribe, once started, had sudden misgivings about the order of selections and had consulted the index, which confirmed the numbering, 1 and 2, and showed the correct position in the second half of the *izbornik*.

A previous deliberate decision to place the initial selections from the second half directly after the answer to Question 20 because of thematic relevance understandably might also have been abandoned eventually as unwise in view of the contradiction with the new index and probably also the text numbering of the second half.

The duplication was probably due to a MS. in 1073's background and not to the scribe who copied this section of 1073. I base this conjecture on the fact that, although the texts are virtually identical (even to the nasal vowels), the distribution of the material in the lines is quite different.

All the borrowed selections come from a single oration of Gregory the Theologian (more commonly, "Nazianzenus"). This is the 40th oration ("In Sanctum Baptisma"). The excerpts in 1073 correspond as follows to the divisions in Migne ([1885], XXXVI, columns 373-385): 151d23ff. — Section 14, fourth sentence through the end, and Section 15, first two sentences; 152c6ff. — Section 11, penultimate of a series of clauses in the final sentence each beginning "ἕως ..." (*donъdeže*); 152c18ff. — Section 12, final sentence; 152c26ff. — Section 19, middle of the fourth sentence through the end, and Section 20, first nine sentences.

One of their notable linguistic features is that they contain two of 1073's three examples of a future participle of *byti*: *byšašt-* 152a1, *byšašč-* 152b8/9.[16] An important 11th century Russian CS version of the compilation of orations called "XIII slovъ Grigorija Bogoslova" (henceforth, XIII), according to Jagić (Kuznecov, p. 129), has nine examples of the participle. Two of them occur in passages corresponding to 1073's excerpts. The third *slovo* in XIII is the same 40th oration. The 1073 excerpts are found, respectively, on the following pages of Anton Budilovič's edition:[17] 82-83 (108a-109a of the MS.); 80 (105c-105d); 81 (106c); 86-87 (113b-115a). 1073's third example, *byšašt-* 159c24/25, is not from a selection by Gregory.

Kuznecov has compiled many more examples of this stem than were previously mentioned in the literature — or perhaps known. A large number come from several 15th and 16th century copies of a now lost Novgorod MS. dated 1047 and signed by the scribe named "Upyrь Lixoj". According to Kuznecov (p. 131), the correspondences between the various copies, both in general and in the shape of the future participle, are so close that one must assume that the 11th century original contained the same forms. Moreover, all the examples so consistently avoid the East Slavic reflex of *tj that they must have been taken from the South Slavic originals, even though the participle is found only in Russian CS MSS. (p. 131). However, one thing remains clear in all this. Though the group of known or conjectured Russian CS MSS. which exhibit the participle is fairly varied, it is, nevertheless, very small. Kuznecov himself makes the same point (p. 129), and he emphasizes also that the participle survived only in certain South Slavic dialects in the 10th and 11th centuries (p. 132). The examples of the future participle in 1073 might, therefore, seem unusual. First, if 1073 genuinely belonged to the limited tradition of MSS. which exhibited the forms, we could expect it to show more examples than it does; such a lexical item has many potential uses. Second, it is curious that a MS. like 1073, so central to the major Kievan scribal and literary tradition, should contain a feature which was otherwise found only in a handful of MSS. — in quantity really only in a single Novgorod MS. But the obvious implication of the absence of the

[16] For a detailed treatment and a review of the literature concerning this residue of a sigmatic future stem, see Petr S. Kuznecov, *Očerki po morfologii praslavjanskogo jazyka* (Moscow, 1961), pp. 127-132.

[17] Same title (Saint Petersburg, 1875).

selections in R^1 is supported by fairly clear evidence that the selections are derivative (and probably recent) in the layers of MSS. underlying 1073 and K since they seem to represent the very same translation or later copy as do the corresponding portions of XIII. The full evidence for this will be adduced in a separate paper. And in view of the unbroken form in XIII, the excerpted form in 1073 must be viewed as secondary. The third example of the future participle in 1073 also appears in R^1. It may have a similar explanation. For the moment, however, it remains the sole example which is potentially original in the MS. tradition represented by 1073.

The nature of an apparent intervention between the original Slavic translation of Gregory's 40th oration and 1073 and K presents something of a puzzle. At the moment, I know of no evidence of another *izbornik* type of Slavic text which contains precisely the combination of material represented by the borrowed selections. In the absence of such evidence, it might seem preferable to assume that the selection and interpolation in the recent background of 1073 and K were done from XIII or an unexcerpted text similar to it and have nothing to do with an *izbornik* consisting at least partly of short excerpts from Gregory.

Nevertheless, there is some indication that the scribe who made the additions extracted the pieces from a source which did contain excerpts. Although the four selections occur in sequence in 1073, K place the last one at the beginning of the second half (unnumbered). Were it not for this, we might be able to blame a temporary oversight for the fact that K is missing the third, short piece and abbreviates the second, also short, by about two lines (after the 10th letter in 152c14 in 1073). Furthermore, the missing portion is much shorter than a column or a page and could not have been missed in the accidental turning of two pages of the protograph. I cannot guess what would have induced a scribe to necessarily place only the last one after the index, unless the four pieces previously had been in *at least* two separate units (of three and one), had then been inserted at separate points in the single source of 1073 and K and had finally been brought together somewhat later in the background of 1073.[18] The headings for the selections in 1073 support this interpretation; whereas the first three form a sequence (*Grigor*[*i*]*a bogoslovьca slovo*, *Togožde* and *Togože*), the fourth repeats the full heading (*Grigoria b*[*og*]*oslovьca slovo*).

[18] Vostokov (p. 501) gives the mistaken impression (possibly inadvertently) that only the last of the four extra selections is not shown in the editions of the Anastasius material provided by Jacob Gretser and Gentian Hervet. I have not been able to find a copy of Hervet's edition. However, I have confirmed Gorskij's explicit statement (p. 373) that none of the four occur in Gretser's edition (*Opera omnia*, IV, Part 2 [Ratisbonae, 1740]). The pertinent section falls between page 277 and page 288, approximately, and has the same contents and ordering as Migne.

There is some strong internal evidence for assigning an approximate date to the original Slavic translation.

First (and most obvious), there is no reason to question the authenticity of the dedication to Simeon in K; clearly, the original translation was made during his reign, that is, between 893 and 927.

The two Greek MSS. are only indirectly useful for dating the translation. Both derive from the early 10th century. The Coislin evidence is quite specific: as Montfaucon notes, a catalogue of the patriarchs of Constantinople "definit in Nicolaum secundo octogesimum nonum, quo sedente scriptus videtur hic Codex, nempe versus annum 912" (p. 194). Unfortunately, this catalogue is not found in the Slavic MSS. Since there probably existed a somewhat earlier version of the Greek *izbornik* from which the Slavic translation could have been made, the Coislin catalogue does not by itself justify moving the early terminus for the translation from 893 to a later date.

However, the catalogue of Byzantine emperors which concludes 1073 and which was mentioned above supplies an approximate date which nearly coincides with 912 and which does justify moving the early terminus. Unlike the preceding entries in this catalogue, the final entry fails to mention the length of the reign in question, although *lět* is written. This reign, then, probably dates the immediate Greek protograph. Contrary to Vostokov (p. 505), the final entry refers not to Constantine IX Monomachus, who reigned from 1042 to 1050 as the third husband of Zoe, daughter of Constantine VIII, as well as for a brief period after her death, but rather to Constantine VII Porphyrogenitus and to his mother, Zoe, who served as regent from 913 to 919, during part of Porphyrogenitus' minority.[19]

This does not rule out the possibility that the Slavic translation was made after 919. We are left, therefore, with a potential span of approximately fifteen years (913-927). Nevertheless, it is doubtful that the literary men of Simeon's politically and culturally vigorous court could have remained unaware for very long of any changes among the regents supervising Porphyrogenitus' early reign. Therefore, 1073's failure to indicate the end of Zoe's regency tends to support 913-920 as the most likely period for the translation.

If we operate with 920 as the approximate date of the OCS original, we are

[19] This catalogue is found in neither of the Greek MSS., although there is some evidence that it may once have formed a part of the Coislin MS. Montfaucon remarks that a catalogue combining the chronologies of a number of royal lines, of which the last is "Imperatorum Romanorum", "mutilus est abscisso folio" (p. 194). Devreesse, on the other hand, does not mention any damage. The Vatican MS. does not supply comparative evidence, because it has none of the chronologies and concludes prior to 1073's last regular selection (*Ipolütovo* ..., 261b21ff.) with the two selections of Pseudo-Epiphanius (Devreesse, 1937, p. 140).

forced to speculate about the century and a half which separates it from 1073 and about the intervening copies. First, let us consider an additional factor about the dating of 1073. As mentioned above, it is based on the signature of the scribe. Gorskij and Nevostruev (pp. 367, 390) raise an issue which would seem to provide additional confirmation of the date but which seems to have been neglected in the literature on 1073. This is that in two places at the end of the text where Svjatoslav's name is mentioned, it is written on scraped parchment. It is not clear from the report of Gorskij et al. or from Duvernois' note to the *Čtenija* edition of 1073 (p. 2) whether S. P. Ševyrev, the discoverer of K, was the first to notice the alterations. However, he was apparently the first in print to conclude that they could only be explained by the events of the year 1073, when Svjatoslav and Vsevolod conspired for the throne of the Great Prince against their brother, Izjaslav. I agree with Gorskij and Duvernois that these events must certainly have provided the occasion for the change of name. But I regard it as far too hasty to conclude that the removed name was necessarily Izjaslav's. An attempt will have to be made under the appropriate chemical and lighting conditions to read the altered text.

But still in the realm of speculation, I think it is proper to question whether the conclusion reached by Gorskij and Duvernois can be naturally and logically reconciled with the fact that the scribe's name is also on scraped parchment. It may indeed turn out that the name was not changed, but rather moved to accomodate the name of the prince. To judge by appearances, this was not necessary. Further, if the name of the prince had been Izjaslav's, then the scribe who originally signed 1073 must have been the one whom Izjaslav had ordered to copy 1073. In glossing over the matter of the scribal name, Gorskij and Duvernois are forced to a conclusion which needs some explanation, namely, that princely competition extended to particular scribes in the scriptoria.

The above does not directly affect the dating of the actual writing of 1073. But it does indirectly concern the "linguistic distance" between 1073 and the original translation and the effects on its language of a possibly long succession of Russian manuscripts underlying it. It cannot be excluded that Simeon's name was still on 1073 when Svjatoslav inherited the manuscript. In that event, there would have been some sense in replacing the name of a very early scribe, perhaps even of the original translator, at the same time that Simeon's name was being replaced. It should be recalled that the lineage of K is not very different from that of 1073. This rather increases the chances that it was only with Svjatoslav that there was any thought of altering the signature and dedication, and possibly even that a South Slavic original had appeared quite late on the Russian scene.

In his description of R^1, Vostokov points out (p. 9) that many of the "Bulgarian" words found in R^1 have been replaced in 1073 by equivalent Russian items. He cites *vasnь* (= *prjamь*) and *svěne* (= *kromě*) as examples. It should be made clear, however, that the replacements were by no means consistent. Both *vasnь* and *svěne* occur fairly frequently in 1073. And, as Gorskij notes (pp. 394-396), a considerable number of Bulgarian words are included among the older layers of 1073's lexicon.

Along the same line of minimizing 1073's South Slavic heritage, Duvernois attaches too much importance to his theory that ever since Jaroslav in the first half of the century the *izbornik* of which 1073 is an example had become, so to speak, the bright jewel of wisdom in the crown of the Great Princes of Kiev (pp. 2-3). By itself the theory is reasonable and explains in part why Svjatoslav should have insisted on the inclusion of his name, as well as the addition of a portrait of himself and his family and an extra rendering of the dedicatory poem at the beginning of the MS. It does not, however, support his further argumentation. In claiming that 1073 is an almost pure example of Russian letters in the tradition instituted by Jaroslav I, he uses it to try to account for the statement in the chronicle that Jaroslav "prekladaše otъ Grekъ na slavjanskoe pisьmo" (p. 3). He adduces evidence from the first seventy-four pages of 1073 which adds up to what he calls a "privyčka pisatь po-grečeski", and which is apparently intended to demonstrate that the South Slavic intermediary was trivial in the Russian adoption of Byzantine letters (pp. 3-4). I find the argument somewhat devious and irrelevant. Furthermore, it is simply incorrect; 1073's grammar, for example, exhibits many features which are found elsewhere only in a conspicuously Bulgarian text such as the Suprasliensis. We should not exaggerate the significance of the redaction done under Jaroslav on which Duvernois' thesis is predicated. This redaction need not have markedly altered the linguistic shape in a distinctively Russian way for all subsequent redactions. If the evidence of the first of the two scribes who copied most of 1073 demonstrates anything, it is that there were circumstances in his training which permitted him to duplicate certain non-Russian norms in his original, to retain and expand on others which had been partially obscured and to completely reconstruct still others on the basis of his own understanding. Even in arguing for a long Russian tradition for the *izbornik*, Duvernois nevertheless makes the point that the norms of Russian-Slavonic writing were still not entirely fixed in the 11th century (p. 4). There was much room for continued active influence from South Slavic MSS. and writing practises. In general it is safe to guess that when a cohesive theory of the history of the *izbornik* is finally constructed, it will be discovered that the relationship of 1073 to its South Slavic original is fairly intimate, in influence if not in time.

One more word concerning the original Slavic text of the *izbornik*. It is

Duvernois' opinion (p. 7) that the generally high quality and accuracy of 1073 cannot be reconciled with some of the mistakes found there. He finds himself forced, in fact, to assume a glagolitic South Slavic original, with which the first Russian scribe had great difficulties. He notes that the highest incidence of mistakes which have been corrected the partial *Čtenija* edition by variant readings from other MSS. occurs in the long passages. My impression of the entire text tends to confirm Duvernois' contention about the distribution of corrupt readings. However, I see no strong reason why these must be attributed to the complexities of a glagolitic protograph. Nor do I believe that the mistakes are of such frequency or magnitude as to require us to assume an exceedingly long succession of copies. A brief inspection of the work of the second scribe of 1073, who evidently did far less toward "correcting" his protograph than did the first scribe, should have convinced Duvernois that the text is in relatively good condition. Examples of the grossest kind of inattention should be no surprise in a MS. of this length. Without the Greek original before him, it might be precisely the good scribe who would compound a small error in the protograph by attempting to restore an intelligible reading, as well as leave obvious errors if he entertained the slightest hesitations. An example of the latter is cited by Duvernois (p. 7); *seovъsemь* (for *slovъsemь*), where the scribe even adds over the *o* the breathing mark which is reserved in his system for the second of a string of vowels. But lastly, there is no need to attribute the majority of mistakes to the Russian scribes. Without further study it is wrong to assume, as Duvernois appears to, that the South Slavic text entered the Russian literary tradition in near perfect condition.

From the very discovery of 1073 there had been almost continuous plans to publish the text. Barsov reviews the history and cites much of the pertinent correspondence in his introduction to the partial edition.

Rumjancev's patronage supported the first attempt, and Aleksandr Vostokov, who was also to compile the catalogue of Rumjancev's library, was chosen as editor in 1824. Ratšin's copy was to be used. Vostokov apparently intended to prepare a critical edition, because he happily reports his discovery of R^1 and its usefulness for the forthcoming edition of 1073. He also attempted to obtain a copy of at least a partial Greek source (pp. xi-xv). Unfortunately, the Migne edition of the Greek had not appeared, and the rare Gretser edition was not available in either Moscow or Saint Petersburg. A 10th century MS. in the Synodal Library, which had been listed in a catalogue of that library's Greek holdings, was the only known source available at the time and was laboriously copied for Vostokov on Rumjancev's commission. This Greek text turned out to be somewhat inadequate for the intended work, because fifty-six answers were missing and the ordering of the rest differed radically from that of 1073 and the Gretser edition.

Rumjancev's death in 1826 interrupted the entire project, since his brother, as Barsov notes resentfully, did not share his enthusiasm for antiquities (p. xviii).

Initially, Rumjancev had concealed the discovery of 1073 (Barsov, pp. iii-iv). Although he had taken pains to silence those who knew about the MS. (ironically, its discoverers as well), the rumor of its existence reached the Obščestvo istorii i drevnostej rossijskix (OIDR), which then tried to obtain the scholarly rights to 1073. The attempts were frustrated. However, when Osip Maksimovič Bodjanskij became secretary of the OIDR in 1845, he initiated the second attempt to publish a critical edition, under its auspices (p. xix ff.). Another copy of 1073 was made, partly by Bodjanskij himself, who was also approved as the editor of the edition. There had been some deterioration in 1073 by this time, and Bodjanskij enlisted Vostokov's help in supplying the missing pages from the Rumjancev copy of 1818-1819 (pp. xx-xxi). Vostokov's description of the Rumjancev collection had now been published (1842), and the existence of both R^1 and of the Coislin MS. was known. Bodjanskij obtained access to R^1, and a copy of the Coislin MS. was made for the OIDR. As of Masing's writing, the latter was on deposit in the library of the OIDR (p. 359).

According to a letter quoted by Barsov (pp. xxi-xxii), Bodjanskij had discovered another, probably 15th century copy of the *izbornik* in the Synodal Library. It is not clear whether this was S^2 (1403; Masing, p. 360 ff.). S^2 is so conspicuously dated and signed (Grěšnyi Onъfimъ; Gorskij, p. 408) that it seems unlikely that Bodjanskij would have been so vague about the date (c. 15th century). The somewhat obscure suggestion given by Barsov's mention of S^2 (p. xxii) as though it were a new topic is that the two MSS. were not the same. However, I am unable to guess what Bodjanskij meant otherwise, since all the MSS. seem to be accounted for.

Subsequently, K, which had been brought to light publicly by Ševyrev in 1850, was incorporated into the critical edition (p. xxiii).

Bodjanskij's editorial work was apparently virtually complete by 1855 (pp. xxii-xxiii). However, by the time of his death in 1877, only twenty-three sheets, representing about a quarter of 1073, had been printed, and these had partially deteriorated. In 1881 the OIDR undertook to issue immediately what was printed, with corrections and commentary by Duvernois. This appeared in the cited issues of the *Čtenija* of the OIDR for 1882. Apparently no attempt was made to reprint the original sheets. In both copies that I have examined, the paper, ink and general motley appearance of the pages which contain the *izbornik* text itself are in sharp contrast to those which contain the introductory material.

Funds for the complete bound edition which Duvernois had originally wished to publish were not available at the time. However, he was asked by the OIDR to proceed with the preparation of the rest of the text in anticipation of completing its publication in subsequent issues of the *Čtenija* (p. xxiv). I have no information about the fate of these further efforts.

Finally, it should be noted that the last of the known versions of the Slavic *izbornik* (Masing's B [for Barsov], p. 364ff.) was discovered in the OIDR'S MS. collection, presumably in 1883 (p. xxiv). It is dated 1519 and signed "Grěšnyi Ionica".

In the meanwhile the massive so-called photolithograph edition of 1073 had appeared.[20] In an interesting introduction, Gennadij Karpov, also a member of the OIDR, mentions that the edition consisted of three hundred and sixty copies (p. x). He also reviews the various activities surrounding 1073. According to Karpov, bad luck continued to attend the plans for the publication of the critical edition. It was only in 1879 that Bodjanskij's widow delivered the MS. of the variant readings for 1073 to the OIDR. The MS. of the prepared text, however, was missing (p. viii). Barsov does not mention the fact, perhaps because the MS. may have been found by 1883, although he is ambiguous on this point. Finally, Karpov speculates that Bodjanskij used S^2 ("1402") to supply the missing initial pages of 1073 (p. ix). There is no question that Bodjanskij used S^2 for comparison. He had had a copy made in 1847 (Barsov, p. xxii). However, R^2, Ratšin's 19th century copy, would obviously be closer to 1073. It is odd that Karpov did not refer to Vostokov's correspondence, which was published in 1873 (Barsov, p. xxi).

An adequate edition of 1073 is long overdue. The "photolithograph" edition has too many errors to be useful for some purposes. Šaxmatov's errata catch only the most glaring ones. One hopes that a new edition will be forthcoming, now that Soviet scholars have begun to attend more seriously to the details of the CS contributions to literary Russian.

[20] *Izbornik Velikogo Knjazja Svjatoslava Jaroslaviča 1073 goda* (= LV of the publications of the Obščestvo ljubitelej drevnej Pis'mennosti) (Saint Petersburg, 1880). Detailed errata are supplied by the seventeen-year-old Šaxmatov: "Zur Textkritik des Codex Sviatoslai von J. 1073 nach der photolithographischen Ausgabe", *Archiv für Slavische Philologie*, VI (1882), 590-597.